Springer Theses

Recognizing Outstanding Ph.D. Research

Aims and Scope

The series "Springer Theses" brings together a selection of the very best Ph.D. theses from around the world and across the physical sciences. Each volume has been selected for its scientific excellence by an internationally renowned institute or department. For greater accessibility to non-specialists, the published versions include an extended introduction, as well as a foreword by the student's supervisor explaining the special relevance of the work for the field. The series as a whole provides a valuable resource both for newcomers to the research fields, and for other scientists seeking detailed background information on special questions. Finally, it serves as an accredited documentation of the valuable contributions made by today's younger generation of scientists.

How to Apply

Theses prepared at internationally leading universities and institutes may be nominated for publication by supervisors with the endorsement of the head of department. Each such department is entitled to nominate one thesis per year. Enquiries and requests for a nomination form should be sent, together with the thesis, to Ms. Nathalie Berg (see link on right).

Eligibility Criteria

To qualify for consideration, theses should fulfil the following criteria:

- They must be written in good English.
- The topic should fall within the confines of Chemistry, Physics, Earth Sciences, Engineering and related interdisciplinary fields such as Materials, Nanoscience, Chemical Engineering, Complex Systems and Biophysics.
- The work reported in the thesis must represent a significant scientific advance.
- A maximum of 30% of the text may be a verbatim reproduction from the author's previous publications.
- They must have been examined and passed during the 12 months prior to nomination.
- The theses should have a clearly defined structure including an introduction accessible to new PhD students and scientists not expert in the relevant field.
- After acceptance, theses should be supplemented by a "Supervisor's Foreword" outlining the wider scientific relevance of the work.

Indexed by zbMATH.

Spandan Mondal

Charming Decays of the Higgs, Z, and W Bosons

Development and Deployment of a New Calibration Method for Charm Jet Identification

Doctoral Thesis accepted by
RWTH Aachen University, Aachen, Germany

Springer

Author
Dr. Spandan Mondal
Department of Physics
Brown University
Providence, RI, USA

Supervisor
Prof. Alexander Schmidt
III. Physikalisches Institut A
RWTH Aachen University
Aachen, Germany

ISSN 2190-5053 ISSN 2190-5061 (electronic)
Springer Theses
ISBN 978-3-031-84676-2 ISBN 978-3-031-84677-9 (eBook)
https://doi.org/10.1007/978-3-031-84677-9

This Springer imprint is published by the registered company Springer Nature Switzerland AG
The registered company address is: Gewerbestrasse 11, 6330 Cham, Switzerland

Von der Fakultät für Mathematik, Informatik und Naturwissenschaften der RWTH Aachen University zur Erlangung des akademischen Grades eines Doktors der Naturwissenschaften genehmigte Dissertation

vorgelegt von
Spandan Mondal, M.Sc.
Serampore, India

Supervisor's Foreword

The discovery of the Higgs boson in 2012 at the world's largest particle collider, the Large Hadron Collider at CERN, marked a major milestone for our fundamental understanding of the structure of matter. This discovery confirms the theory that masses of elementary particles are created through their interaction with a scalar quantum field, the Higgs field. While the Higgs boson is considered to be the most recent addition to the completion of the Standard Model of particle physics, the puzzle is not yet complete. Various fundamental problems are still waiting to be clarified, such as the question about the nature of Dark Matter in the universe or the reason why the Higgs boson has a mass of 125 GeV, an apparent result of a mysterious fine tuning mechanism. It is probably the most important task of modern particle physics to look for answers to these questions. The answers are most likely connected to the Higgs boson in one way or another.

One possibility to find answers is the study of the properties of the Higgs boson itself, such as how it couples to other particles. This is the subject of Spandan Mondal's doctoral thesis. In particular, this thesis investigates the question whether the Higgs boson couples to the Charm Quark, how strongly it does that and whether it agrees with the theoretical expectations. While the question sounds simple, the world's biggest and most complex experiments are needed to investigate it. The technical challenges are tremendous both in terms of operating the particle detectors and analysing the incredible amount of data produced by the detectors. Before we can say anything about the Higgs boson, we first need to make sense of the detector data and the detector performance. The concrete challenge in Spandan Mondal's research work is also connected to this task. He developed and deployed a new strategy to identify the signatures of Charm Quarks in the data (so-called Charm Jets) that could be decay products of the Higgs boson. The important aspect here is the calibration of the performance of the Charm Jet identification algorithms. This is necessary to verify that the Charm identification algorithms have the expected efficiencies and failure rates, along with the corresponding systematic uncertainties.

This calibration technique is a fundamental ingredient for the actual search for Higgs bosons decaying to Charm quarks for which Spandan has made decisive contributions as well. It was possible to determine upper limits on the coupling strength between Higgs boson and Charm Quark, dramatically exceeding the expectations for the current amount of data delivered by the LHC. We are therefore now optimistic that the increased amounts of data expected from future LHC runs might lead us to an actual evidence for the Higgs boson decaying to Charm quarks, thus eliminating the much more pessimistic expectations before the release of this result. This might lead us to another major milestone in particle physics in the near future.

Aachen, Germany
January 2025

Prof. Alexander Schmidt

Abstract

This thesis presents searches for the Higgs boson, Z boson, and W boson decaying into charm quark(s) performed with proton-proton collision data at $\sqrt{s} = 13$ TeV collected by the CMS experiment at the LHC, CERN, corresponding to an integrated luminosity of 138 fb^{-1} and recorded between 2016 and 2018. The searches are carried out using events in which the Higgs/Z/W boson is produced in association with a leptonically decaying Z or W boson. This thesis also discusses a novel calibration algorithm for charm jet identification that enables maximal use of the available information related to charm jets. The new method is used to correct the entire distribution expected as output when jet flavour identification algorithms are applied to jets of different flavours. The calibrated results improve over traditional efficiency measurements and help enhance the sensitivities of the Higgs, Z, and W searches.

This thesis primarily reports on the so-called resolved-jet topology of the Higgs/Z/W boson searches, where the boson candidates are reconstructed using two separate small-radius (AK4) jets. Upon statistically combining the results of the resolved-jet search with a complementary merged-jet approach, the observed (expected) upper limit on the Higgs decay to charm quarks corresponds to 14 (7.6) times the Standard Model expectation at the 95% CL which is the most stringent to date. A significance of 5.7σ (5.9σ) over the background-only prediction is observed (expected) in case of the search for charmed Z boson decays. Only the resolved-jet topology is used in the search for charmed W decays and the observed (expected) significance is 5.6σ (5.7σ) over the background-only prediction. These mark the first observations of charmed decays of the Z and W bosons at a hadron collider experiment.

Acknowledgements This work was made possible due to the support of a number of people, both in my professional and personal circles. I would like to express my heartfelt gratitude to them.

Firstly, many thanks to Prof. Alexander Schmidt for being an amazing and supportive supervisor. This work would definitely not be possible without his constant guidance and encouragement over the years. I feel truly lucky to have had an exceptionally supportive doctoral supervisor.

I would also like to thank Prof. Michal Czakon, who agreed to be my second supervisor and one of the referees of my thesis, for his cooperation and advice whenever I needed help.

My groupmates have also been extremely supportive. Dr. Luca Mastrolorenzo, Dr. Xavier Coubez, and Dr. Andrey Pozdnyakov have been excellent mentors who guided me in several aspects of my thesis. My colleagues, Andrzej Novak, Alena Dodonova, Wahid Redjeb, Ming-Yan Lee, Annika Stein, and Alexander Jung were excellent collaborators and the several discussions with them have been very productive. I would particularly like to thank Prof. Schmidt and Andrzej for helping me in proofreading my thesis, and Alena for helping me organise the thesis colloquium.

The Research Training Group (RTG) "Physics of the Heaviest Particles at the LHC" at RWTH provided me with a platform and community to exchange and discuss ideas. I would like to thank all its members for the valuable discussions over the years.

I would like to thank the European Research Council (ERC) for funding this work under the European Union's Horizon 2020 research and innovation programme (Grant Agreement No. 724704). I would also like to thank the Federal Ministry of Education and Research (BMBF), Germany, for funding research travel through the RTG.

I am also grateful to several collaborators at the Compact Muon Solenoid (CMS) experiment, CERN, for productive discussions and fruitful collaborative research. In particular, I would like to thank members of the Hbb subgroup and BTV group in CMS for their support.

The Physics Institute III A and the RWTH Aachen University provided me with an excellent research environment, an unparalleled world-class work culture, and an opportunity to perform this research work. I am grateful to the institute and university members for that.

Finally, I would like to thank my family and friends for their unconditional support. None of this would have been possible without them being by my side through the years.

Competing Interests The author has no competing interests to declare that are relevant to the content of this manuscript.

Contents

Chapter 1
Introduction

The observation of the Higgs boson by the ATLAS [1] and CMS [2, 3] experiments at the Large Hadron Collider (LHC), CERN, in 2012 marked a giant leap in mankind's extensive efforts towards understanding the fundamental constituents of matter and the forces that act on them. This was, however, only the beginning of a new series of exciting searches in the field of experimental particle physics. Over the next decade, a tremendous volume of research went into understanding the properties of the particle, all of which has culminated in an improved understanding of the Higgs boson as well as its interactions with other fundamental particles.

The Standard Model (SM) of particle physics is an elaborate theory that describes the fundamental interactions between elementary particles. It predicts the existence of the Higgs boson as a consequence of the so-called Higgs mechanism [4–6], which also explains the origin of masses of force-carrying particles (bosons). In addition, the Yukawa interaction in the SM describes the interaction of the Higgs boson with matter particles (fermions) and explains the origin of masses of the latter. Thus, fermions in the SM assume a mass proportional to the strength of their Yukawa coupling to the Higgs boson.

While the SM has survived a vast majority of experimental verifications so far, certain observations are not explained by the SM. Examples of such observations include evidence of the existence of dark matter [7], neutrino flavour oscillations [8–10], matter-antimatter asymmetry [11], etc. These point towards the existence of more complete theories, often referred to as theories beyond the SM (BSM).

A decade of hadron collider searches has firmly established the coupling of the Higgs boson to the gauge bosons (Z [12, 13], W [14–16], and photon [17, 18]) as well as to the heavy fermions (top [19], bottom quarks [20, 21], tau leptons [22–24]). In these searches, no deviations from the SM predictions have been observed. While there is recent evidence of the Higgs boson coupling to the muon [25], our current experiments are not sensitive enough to observe the Higgs boson coupling to the charm, strange, up, down quarks, or the electron. Hence, the next milestone of the

S. Mondal, *Charming Decays of the Higgs, Z, and W Bosons*, Springer Theses,
https://doi.org/10.1007/978-3-031-84677-9_1

LHC physics programme is to establish Higgs couplings to the second-generation quarks (charm and strange). The previous direct searches for the Higgs boson decaying to charm quarks at the LHC by the ATLAS [26, 27] and CMS [28] collaborations target the associated production of a Higgs boson with a vector boson (V). The most recent direct search by the ATLAS collaboration obtained an upper limit on the VH($H \rightarrow c\bar{c}$) process of 26 times the SM prediction at the 95% confidence level [27]. A more recent result from the ATLAS collaboration combines results from the direct search with those from differential Higgs cross section measurements, providing a more stringent, indirect upper limit on the Higgs–charm coupling [29].

Since the Yukawa couplings of lighter fermions are weaker compared to those of heavier ones, light-fermion couplings could potentially be significantly altered by the presence of additional (or alternative) BSM coupling terms [30–33], and hence measurements sensitive to such light couplings are also likely to be sensitive towards BSM coupling terms that potentially complement or replace the SM Yukawa coupling. Thus, besides serving as an important crosscheck of SM predictions, searches for light-fermion couplings are also a promising gateway to search for BSM physics.

This thesis reports on the most sensitive direct search of the Higgs boson coupling to charm quarks to date, in events that contain a Higgs boson decaying to a charm quark-antiquark pair and produced in association with a leptonically decaying W or Z boson ("VHcc" search). The search has been carried out using data collected by the CMS experiment corresponding to 138 fb^{-1} of proton-proton (pp) collision events with a centre of mass energy ($\sqrt{s}$) of 13 TeV. The general strategy is to search for events with one or two charged leptons or missing energy in the final state, along with at least two small-radius jets that are likely to have originated from charm quarks. Such a search is especially challenging due to the small branching fraction of the Higgs boson to charm quarks, as well as due to complexities involved in identifying charm-quark initiated jets. Several modern analysis techniques have been implemented to optimise the search. A new calibration method of charm jet identification algorithms has been developed and published as Ref. [34]. In combination with an alternative strategy to reconstruct large-radius merged jets, the results of this search have been published in Ref. [35]. This thesis and Ref. [35] also report the observation of the Z boson decaying to a charm quark-antiquark pair, marking the first observation of such a process at a hadron collider experiment ("VZcc" search). A similar analysis using a subset of this data corresponding to 35.9 fb^{-1} was also published earlier as Ref. [28]. This thesis additionally reports on the observation of the W boson decaying to a pair of quarks, one of which is a charm quark ("VWcq" search). This also marks the first observation of charmed W boson decays in diboson events at a hadron collider experiment.

The thesis is structured in 11 sections. Sections 2 and 3 discuss the theoretical and experimental backgrounds, respectively. Section 4 outlines the technologies implemented for efficient data analysis. Section 5 describes the selection criteria used to define reconstructed physics objects for the analyses. Sections 6 and 7 discuss the algorithms used for identification of the flavour of the jets and well as the novel method used to calibrate such algorithms. The analysis strategy is described in Sect. 8 and the details of the analyses are explained in Sect. 9. The results of the statistical analysis of the data are summarised in Sect. 10. Finally, a conclusion of the thesis is presented in Sect. 11.

References

1. ATLAS Collaboration (2012) Observation of a new particle in the search for the Standard Model Higgs boson with the ATLAS detector at the LHC. Phys Lett B 716:1. https://doi.org/10.1016/j.physletb.2012.08.020. arXiv:1207.7214
2. CMS Collaboration (2012) Observation of a new boson at a mass of 125 GeV with the CMS experiment at the LHC. Phys Lett B 716:30–61. https://doi.org/10.1016/j.physletb.2012.08.021. arXiv:1207.7235
3. CMS Collaboration (2013) Observation of a new boson with mass near 125 GeV in pp collisions at $\sqrt{s} = 7$ and 8 TeV. JHEP 06:081. https://doi.org/10.1007/JHEP06(2013)081. arXiv:1303.4571
4. Englert F, Brout R (1964) Broken symmetry and the mass of gauge vector mesons. Phys Rev Lett 13:321. https://doi.org/10.1103/PhysRevLett.13.321
5. Higgs PW (1964) Broken symmetries, massless particles and gauge fields. Phys Lett 12:132–133. https://doi.org/10.1016/0031-9163(64)91136-9
6. Higgs PW (1964) Broken symmetries and the masses of gauge bosons. Phys Rev Lett 13:508–509. https://doi.org/10.1103/PhysRevLett.13.508
7. Trimble V (1987) Existence and nature of dark matter in the universe. Annu. Rev. Astron. Astrophys. 25:425–472. https://doi.org/10.1146/annurev.aa.25.090187.002233
8. Davis R Jr, Harmer DS, Hoffman KC (1968) Search for neutrinos from the sun. Phys Rev Lett 20:1205–1209. https://doi.org/10.1103/PhysRevLett.20.1205
9. SNO Collaboration (2001) Measurement of the rate of $\nu_e + d \rightarrow p + p + e^-$ interactions produced by ^{8}B solar neutrinos at the sudbury neutrino observatory. Phys Rev Lett 87:071301. https://doi.org/10.1103/PhysRevLett.87.071301. arXiv:nucl-ex/0106015
10. Super-Kamiokande Collaboration (1998) Evidence for oscillation of atmospheric neutrinos. Phys Rev Lett 81:1562–1567. https://doi.org/10.1103/PhysRevLett.81.1562. arXiv:hep-ex/9807003
11. Sakharov AD (1967) Violation of CP Invariance, C asymmetry, and baryon asymmetry of the universe. Pisma Zh Eksp Teor Fiz 5:32–35. https://doi.org/10.1070/PU1991v034n05ABEH002497
12. CMS Collaboration (2014) Measurement of the properties of a Higgs boson in the four-lepton final state. Phys Rev D 89:092007. https://doi.org/10.1103/PhysRevD.89.092007. arXiv:1312.5353
13. ATLAS Collaboration (2015) Measurements of Higgs boson production and couplings in the four-lepton channel in pp collisions at center-of-mass energies of 7 and 8 TeV with the ATLAS detector. Phys Rev D 91:012006. https://doi.org/10.1103/PhysRevD.91.012006. arXiv:1408.5191
14. CMS Collaboration (2014) Measurement of Higgs boson production and properties in the WW decay channel with leptonic final states. JHEP 01:096. https://doi.org/10.1007/JHEP01(2014)096. arXiv:1312.1129
15. ATLAS Collaboration (2015) Observation and measurement of Higgs boson decays to WW* with the ATLAS detector. Phys Rev D 92:012006. https://doi.org/10.1103/PhysRevD.92.012006. arXiv:1412.2641
16. ATLAS Collaboration (2015) Study of (W/Z)H production and Higgs boson couplings using H $\rightarrow$ WW* decays with the ATLAS detector. JHEP 08:137. https://doi.org/10.1007/JHEP08(2015)137. arXiv:1506.06641

17. CMS Collaboration (2014) Observation of the diphoton decay of the Higgs boson and measurement of its properties. Eur Phys J C 74:3076. https://doi.org/10.1140/epjc/s10052-014-3076-z. arXiv:1407.0558
18. ATLAS Collaboration (2014) Measurement of Higgs boson production in the diphoton decay channel in pp collisions at center-of-mass energies of 7 and 8 TeV with the ATLAS detector. Phys Rev D 90:112015. https://doi.org/10.1103/PhysRevD.90.112015. arXiv:1408.7084
19. ATLAS Collaboration (2018) Observation of Higgs boson production in association with a top quark pair at the LHC with the ATLAS detector. Phys Lett B 784:173. https://doi.org/10.1016/j.physletb.2018.07.035. arXiv:1806.00425
20. CMS Collaboration (2018) Observation of Higgs boson decay to bottom quarks. Phys Rev Lett 121:121801. https://doi.org/10.1103/PhysRevLett.121.121801. arXiv:1808.08242
21. ATLAS Collaboration (2018) Observation of $H \to b\bar{b}$ decays and VH production with the ATLAS detector. Phys Lett B 786:59. https://doi.org/10.1016/j.physletb.2018.09.013. arXiv:1808.08238
22. CMS Collaboration (2014) Evidence for the 125 GeV Higgs boson decaying to a pair of τ leptons. JHEP 05:104. https://doi.org/10.1007/JHEP05(2014)104. arXiv:1401.5041
23. ATLAS Collaboration (2015) Evidence for the Higgs-boson Yukawa coupling to tau leptons with the ATLAS detector. JHEP 04:117. https://doi.org/10.1007/JHEP04(2015)117. arXiv:1501.04943
24. CMS Collaboration (2017) Observation of the Higgs boson decay to a pair of tau leptons. Phys Lett B 779:283. https://doi.org/10.1016/j.physletb.2018.02.004. arXiv:1708.00373
25. CMS Collaboration (2021) Evidence for Higgs boson decay to a pair of muons. JHEP 01:148. https://doi.org/10.1007/JHEP01(2021)148. arXiv:2009.04363
26. ATLAS Collaboration (2018) Search for the decay of the Higgs boson to charm quarks with the ATLAS experiment. Phys Rev Lett 120:211802. https://doi.org/10.1103/PhysRevLett.120.211802. arXiv:1802.04329
27. ATLAS Collaboration (2022) Direct constraint on the Higgs-charm coupling from a search for Higgs boson decays into charm quarks with the ATLAS detector. Eur Phys J C 82:717. https://doi.org/10.1140/epjc/s10052-022-10588-3. arXiv:2201.11428. Submitted to EPJC
28. CMS Collaboration (2020) A search for the standard model Higgs boson decaying to charm quarks. JHEP 03:131. https://doi.org/10.1007/JHEP03(2020)131. arXiv:1912.01662
29. ATLAS Collaboration (2023) Measurement of the total and differential Higgs boson production cross-sections at $\sqrt{s}$ = 13 TeV with the ATLAS detector by combining the H $\to$ ZZ* $\to$ 4ℓ and H $\to \gamma\gamma$ decay channels. JHEP 05:028. https://doi.org/10.1007/JHEP05(2023)028. arXiv:2207.08615
30. Ghosh D, Gupta RS, Perez G (2016) Is the Higgs mechanism of fermion mass generation a fact? A Yukawa-less first-two-generation model. Phys Lett B 755:504. https://doi.org/10.1016/j.physletb.2016.02.059. arXiv:1508.01501
31. Botella FJ, Branco GC, Rebelo MN, Silva-Marcos JI (2016) What if the masses of the first two quark families are not generated by the standard model Higgs boson? Phys Rev D 94:115031. https://doi.org/10.1103/PhysRevD.94.115031. arXiv:1602.08011
32. Harnik R, Kopp J, Zupan J (2013) Flavor violating Higgs decays. JHEP 03:026. https://doi.org/10.1007/JHEP03(2013)026. arXiv:1209.1397
33. Altmannshofer W et al (2016) Collider signatures of flavorful Higgs bosons. Phys Rev D 94:115032. https://doi.org/10.1103/PhysRevD.94.115032. arXiv:1610.02398
34. CMS Collaboration (2022) A new calibration method for charm jet identification validated with proton-proton collision events at $\sqrt{s}$ =13 TeV. JINST 17:P03014. https://doi.org/10.1088/1748-0221/17/03/P03014. arXiv:2111.03027
35. CMS Collaboration Search for Higgs boson decay to a charm quark-antiquark pair in proton-proton collisions at $\sqrt{s}$ = 13 TeV. Phys Rev Lett 131:061801. https://doi.org/10.1103/PhysRevLett.131.061801. https://doi.org/10.48550/ARXIV.2205.05550. arXiv:2205.05550. Accepted for publication by PRL

Chapter 2
The Standard Model

This section presents a brief overview of the Standard Model (SM) of particle physics, with focus on the concepts relevant to the analyses presented in this thesis.

2.1 The Standard Model of Particle Physics

Our current understanding of the elementary particles of nature and the interactions between them can be summarised with the SM of particle physics. The SM describes three of the four known fundamental forces of nature—electromagnetic (EM), weak, and strong interactions—under the interpretation of these as quantised fields. Gravity, the fourth force, is not described by the SM and can be considered negligible for all practical purposes at the relevant energy scales (of the order of several TeVs, and lower).

The SM describes dynamics of particles as interactions of quantum gauge field theories [1–4]. Particles are represented as quantum fields and their dynamics are described by the Lagrangian density, or simply the Lagrangian, $\mathcal{L}$. The form of the Lagrangian is constrained using symmetries that the fields follow, and the Lagrangian remains invariant under transformations of the governing symmetry group [5]. The symmetries imposed on the particle fields follow those of the Poincare group [6, 7], implying conservation of energy, momentum, and angular momentum.

The SM particles can be classified into two major categories: fermions and bosons, depending on whether a particle follows Fermi-Dirac (spin $\frac{1}{2}$) [8, 9] or Bose-Einstein (integral spin) [10] statistics, respectively. Figure 2.1 summarises the particles and the following subsections discuss each category.

S. Mondal, *Charming Decays of the Higgs, Z, and W Bosons*, Springer Theses,
https://doi.org/10.1007/978-3-031-84677-9_2

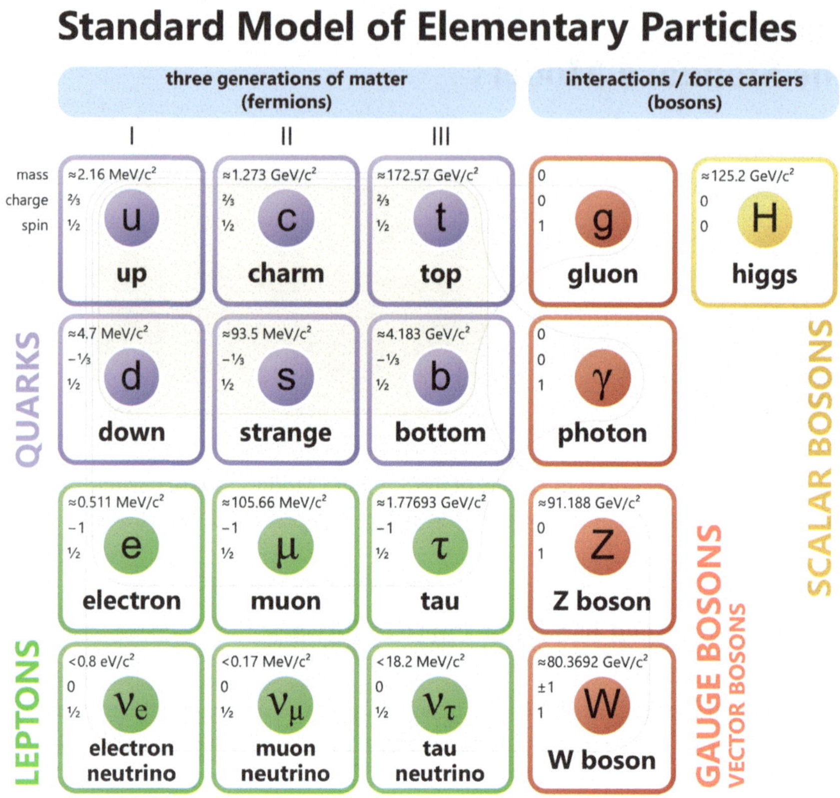

Fig. 2.1 Standard model of particle physics: the 12 fermions and 5 bosons. Brown loops indicate which bosons (red) couple to which fermions (purple and green) [11, 12]

2.1.1 Fermions

Fermions are elementary matter particles with spin $\frac{1}{2}$. They can be divided into two categories: quarks (u, d, c, s, t, b) and leptons (e, ν_e, μ, ν_μ, τ, ν_τ), and their corresponding antiparticles. They can be divided into three generations as shown in Table 2.1.

Quarks with positive charges $Q = +\frac{2}{3}$ (in units of the charge of the electron) are called "up"-type and ones with negative charges $Q = -\frac{1}{3}$ are called "down"-type. They also possess the so-called colour charge and weak isospin, and interact with the strong, weak, and EM forces. Quarks do not exist freely in nature; they always form colour-neutral bound states of either a quark-antiquark pair called mesons, or a trivalent bound state called baryons. Exotic bound states such as tetra- and pentavalent quarks have also been observed more recently [13, 14]. Together, such bound states are called hadrons. Only the top quark decays faster

Table 2.1 The fermions in the SM of particle physics with their EM charge and weak isospin values. Neutrinos are denoted with ν. Each type of quark further has three types of colour charges. These particles along with their corresponding antiparticles constitute the fermions of the SM

Fermion type		Generation			Charge	Weak isospin
		I	II	III		
Quarks	"Up"-type	Up	Charm	Top	$+\frac{2}{3}$	$+\frac{1}{2}$
	"Down"-type	Down	Strange	Bottom	$-\frac{1}{3}$	$-\frac{1}{2}$
Leptons	Charged	Electron	Muon	Tau	-1	$-\frac{1}{2}$
	Neutral	ν_e	ν_μ	ν_τ	0	$+\frac{1}{2}$

($\sim 10^{-25}$ s) than the hadronisation timescale ($\sim 10^{-24}$ s) and therefore does not form any hadrons. As discussed in upcoming subsections in context of the strong interaction, quarks generated in the hard scattering processes at hadron colliders induce a chain of hadronisations resulting in a collimated stream of particles called jets. Thus, quarks (with the exception of the top quark) are detected in their final states as jets, while the top quark can be reconstructed from its decay products.

Leptons can be further classified as charged and neutral leptons. Charged leptons (e, μ, τ) have a charge of $Q = -1$ and can interact electromagnetically. Neutral leptons, on the other hand, are called neutrinos (ν_e, ν_μ, ν_τ) and do not have an electric charge. Both charged and neutral leptons possess a weak isospin and hence can interact with the weak force. Although the neutrinos in the SM are massless, recent experimental evidence of neutrino flavour oscillations show that they, in fact, have non-zero masses [15–17]. The electron is a stable particle, while the muon and tau have lifetimes of the order of $\sim 10^{-6}$ s and $\sim 10^{-13}$ s, respectively, in their rest frames. The muon can, however, traverse entire detectors when produced at high energies, and hence can be identified directly at high-energy hadron colliders. Taus, on the other hand, can be detected only through their leptonic and hadronic decay products.

2.1.2 *Bosons*

Bosons are force-mediating particles in the SM. All fundamental interactions in the SM are described as a quantum field theory. Requiring local gauge invariance (cf. Sect. 2.2) of these quantum fields results in the recovery of the known interactions: quantum chromodynamics (QCD) for the strong force, the weak interactions for the weak force, and quantum electrodynamics (QED) for the electromagnetic force. The excitation modes of these gauge fields (after spontaneous symmetry breaking and mixing, cf. Sect. 2.3.1) are referred to as gauge bosons and have an integer spin of 1. The gauge bosons of the SM are listed in Table 2.2.

The gauge fields corresponding to the three forces do not carry any mass terms in the Lagrangian, as introduction of mass terms violates gauge invariance. Hence,

Table 2.2 List of gauge bosons in the SM along with their respective electric charges and masses. Gluons can further have 8 different colour charges

Boson	Charge	Mass (GeV)	Mediator of...
Photon (γ)	0	0	EM
Gluon (g)	0	0	Strong
W^+	+1	80.379 ± 0.012	Weak
W^-	−1		Weak
Z	0	91.188 ± 0.002	Weak

the discovery of the massive W and Z bosons [18, 19] was historically at odds with the requirements of gauge invariance. This apparent conflict was resolved with the introduction of a spontaneous symmetry breaking mechanism (cf. Sect. 2.3.1), and hence the addition of a complex scalar field, called the Higgs field. As the Higgs field is a quantum field itself, it gives rise to an additional boson, called the Higgs boson, which has a spin 0 and properties different from those of gauge bosons. The Higgs boson along with the spontaneous symmetry breaking mechanism is discussed in more detail in Sect. 2.3.

2.2 Interactions

2.2.1 Overview

In addition to the global Poincare symmetry that accounts for conservation of energy, momentum, and angular momentum, SM particle fields follow a local symmetry under the unitary group $SU(3) \times SU(2) \times U(1)$. These correspond to three symmetries described as follows:

- SU(3): Conservation of colour charge under interactions with gauge field G,
- SU(2): Conservation of weak isospin charge under interactions with gauge field W,
- U(1): Conservation of weak hypercharge under interactions with gauge field B.

However, requiring local gauge invariance results in the fields being massless, which is at odds with experimental measurements of massive Z and W bosons. This apparent conflict is resolved by unifying the weak and QED interactions into a unified electroweak (EW) gauge group represented by $SU(2) \times U(1)$ [20], which is spontaneously broken into only one $U(1)_{QED}$ symmetry with the Higgs mechanism. The Y and B states after the symmetry breaking mechanism mix to yield the massive $W^{\pm}$ and Z bosons, and a massless photon (γ). It is to be noted that this mixing results in the post-symmetry-breaking $U(1)_{QED}$ being different from the pre-symmetry-breaking U(1). The Higgs field, on the other hand, is a doublet of the SU(2) group. As the original SU(2) × U(1) symmetry is broken, the weak isospin and weak hypercharge are no longer conserved in general, and instead an "electric charge", corresponding to the post-symmetry-breaking $U(1)_{QED}$ symmetry, is conserved.

The field G (with the SU(3) symmetry) corresponds to the QCD gauge field, which remains unaffected by the Higgs mechanism. This is consistent with the gluon being massless. This group being unaffected by the Higgs mechanism also results in the colour charge being conserved even after spontaneous symmetry breaking. QCD (SU(3)) can therefore be combined independently with the EW group (SU(2) × U(1)) to form the full SM gauge group SU(3) × SU(2) × U(1), which is different from the initial SU(3) × SU(2) × U(1) described in this subsection, and accounts for massive bosons as well.

The following subsections contain brief descriptions of the strong, electromagnetic, and weak interactions, followed by that of electroweak unification.

2.2.2 *Strong Interaction*

QCD is a non-abelian gauge theory based on the local gauge symmetry group SU(3) and governs the strong interaction. The QCD group has three orthogonal charge quantum numbers associated to it, usually denoted by red, green, and blue. These quantum numbers are often referred to as colour charges.

The free-field Lagrangian of a quark with a quark field q and mass m is given by the Dirac Lagrangian

$$\mathcal{L} = \bar{q}(i\gamma^{\mu}\partial_{\mu} - m)q, \tag{2.1}$$

where γ^{μ} represent the Dirac matrices and ∂_{μ} represents the partial derivative. The transformation of q under an SU(3) transformation can be given by

$$q \rightarrow q' = Uq, \tag{2.2}$$

where U is a 3 × 3 unitary matrix of the form

$$U = e^{ig_s \frac{\lambda^a}{2} \alpha_a}, \tag{2.3}$$

g_s is a real constant, α_a represent the parameters of the transformation, and $\frac{\lambda^a}{2}$ denote the 8 Gell-Mann matrices, corresponding to the generators of the SU(3) group. For the Lagrangian $\mathcal{L}$ to be invariant under this transformation, the derivative ∂_{μ} has to be redefined to the covariant derivative given by

$$D_{\mu} = \partial_{\mu} + ig_s \frac{\lambda_a}{2} G^a_{\mu}, \tag{2.4}$$

where G^a_μ are vector boson fields that can be identified as the eight gluon fields that transform as

$$G^a_\mu \rightarrow G'^a_\mu = G^a_\mu + \partial_\mu \alpha^a + g_s f_{abc} \alpha^b G^c_\mu, \tag{2.5}$$

where f_{abc} are the structure constants of the SU(3) group and satisfy the commutation rules $\left[\frac{\lambda^a}{2}, \frac{\lambda^b}{2}\right] = i f_{abc} \frac{\lambda^c}{2}$.

An additional kinetic energy term is introduced in the Lagrangian for the propagation of the gluon field and is given by $-\frac{1}{4} G^{\mu\nu}_a G^a_{\mu\nu}$, where $G^a_{\mu\nu}$ is the gluon field strength expressed as a tensor of the form

$$G^a_{\mu\nu} = \partial_\mu G^a_\nu - \partial_\nu G^a_\mu - g_s f_{abc} G^b_\mu G^c_\nu. \tag{2.6}$$

Substituting these terms in the Lagrangian in Eq. 2.1 gives

$$\mathcal{L}_{\text{QCD}} = i\bar{q}\gamma^\mu \partial_\mu q - m\bar{q}q - g_s \bar{q}\gamma^\mu \frac{\lambda_a}{2} q G^a_\mu - \frac{1}{4} G^{\mu\nu}_a G^a_{\mu\nu}, \tag{2.7}$$

which is the QCD Lagrangian invariant under local SU(3) transformations. The first term in the Lagrangian represents the free-field propagation of q. The second term is the mass term. The third term describes the interaction of the gluon with a quark. The coupling constant g_s determines the strength of this interaction and it is often redefined as

$$\alpha_s = \frac{g_s^2}{4\pi}, \tag{2.8}$$

known as the strong coupling constant. The fourth term represents propagation of gluons. As this term is not linear in terms of gluon potential, it introduces trilinear and quadrilinear terms in the Lagrangian, corresponding to 3-gluon and 4-gluon self-interactions. It can be noted that the Lagrangian does not contain a mass (m_g) term $m_g^2 G^\mu_a G^a_\mu$ for the gluons, and therefore results in the gluons being massless in the SM.

2.2.3 Electromagnetic Interaction

The electromagnetic interaction is represented with the QED Lagrangian. QED is an abelian gauge theory with the symmetry group $\text{U}(1)_{\text{QED}}$ and explains interactions of electrically charged particles. With the same approach as in QCD, the QED Lagrangian can be expressed as

$$\mathcal{L}_{\text{QED}} = i\bar{\psi}\gamma^\mu \partial_\mu \psi - m\bar{\psi}\psi - eQ\bar{\psi}\gamma^\mu \psi A_\mu - \frac{1}{4} F_{\mu\nu} F^{\mu\nu}, \tag{2.9}$$

with ψ as a fermion field, e as the charge of the electron, Q as the quantum number associated to QED, and $F_{\mu\nu}$ as the electromagnetic or Maxwell tensor given by

$$F_{\mu\nu} = \partial_\mu A_\nu - \partial_\nu A_\mu. \tag{2.10}$$

The first term in the Lagrangian in Eq. 2.9 represents the free-field propagation of ψ, while the second term is the mass term, with m as the mass of the fermion. The third term arises as a result of the covariant derivative given by

$$D_\mu = \partial_\mu + ieQA_\mu, \tag{2.11}$$

and represents the interaction between the gauge potential A_μ and the fermion field. In case of QED, A_μ represents the photon (γ). The last term in the Lagrangian corresponds to the free propagation of the photon. Similar to QCD, the QED Lagrangian does not contain a mass term for the photon and hence the photon is massless. However, the absence of higher order terms in Eq. 2.10 results in the photon not interacting with itself, in contrast to the gluon field.

2.2.4 *Weak Interaction*

The theoretical description of the weak interaction is more complex, as experimental observations [21] conclude that it violates parity. This is accounted for by assigning different interaction terms to fermions with different "chiralities". This is achieved by using the gamma matrix $\gamma^5 = i\gamma^0\gamma^1\gamma^2\gamma^3$, whose eingenvalues can be either -1 or $+1$. This gives rise to "left" (ψ_L) and "right" (ψ_R) chirality fields. Projection operators, P_L and P_R, are defined as

$$P_L = \frac{1-\gamma^5}{2};\ P_R = \frac{1+\gamma^5}{2}, \tag{2.12}$$

such that for a pair of fermion fields, (ψ,ψ'), one obtains the left and right chirality fields with the operations,

$$\begin{aligned}\Psi_L &= P_L \begin{pmatrix}\psi\\ \psi'\end{pmatrix} = \begin{pmatrix}\psi_L\\ \psi'_L\end{pmatrix},\\ \psi_R &= P_R\psi,\\ \psi'_R &= P_R\psi',\end{aligned} \tag{2.13}$$

where Ψ_L is a left-chirality SU(2) doublet and (ψ_R, ψ'_R) are two right-chirality SU(2) singlets. The (ψ, ψ') pair represents either a (ν, l) pair or an (u, d) pair, where ν, l, u, and d represent a neutrino, a charged-lepton, an up-type quark, and a down-type quark, respectively, in the same generation of fermions.

The Lagrangian for the weak interaction, with the omission of the mass term, can be written as

$$\mathcal{L}_{\text{weak}} = i\bar{\Psi}_L\gamma^\mu D_\mu \Psi_L + i\bar{\psi}_R\gamma^\mu D_\mu \psi_R + i\bar{\psi'}_R\gamma^\mu D_\mu \psi'_R - \frac{1}{4}W^i_{\mu\nu}W^{\mu\nu}_i, \tag{2.14}$$

where $W^i_{\mu\nu}$ is a tensor field defined as

$$W^i_{\mu\nu} = \partial_\mu W^i_\nu - \partial_\nu W^i_\mu - g_w \epsilon_{ijk} W^j_\mu W^k_\nu, \tag{2.15}$$

with g_w as the weak coupling constant, ϵ_{ijk} as the Levi-Civita tensor, and W^i_μ ($i = 1, 2, 3$) as three gauge fields. The covariant derivative in Eq. 2.14 is given by

$$D_\mu = \partial_\mu + i g_w T^i_\mu W^i_\mu, \tag{2.16}$$

where $T^i_\mu = \frac{\sigma_i}{2}$ are the generators of the SU(2) group, proportional to the Pauli matrices, σ_i. It is to be noted that the W^i_μ boson fields do not correspond to the physical $W^\pm$ and Z bosons. As in the case of QCD, the higher-order terms in Eq. 2.15 result in boson self-interactions in the weak interaction.

The quantum number associated to the SU(2) field of the weak interaction is called the weak isospin. It has three components, denoted by I_1, I_2, and I_3. The third component, I_3, is zero for right-chirality fermion fields, ψ_R and ψ'_R. The left-chirality field, Ψ_L has I_3 components $\begin{pmatrix} +1/2 \\ -1/2 \end{pmatrix}$, as it forms a doublet of the SU(2) group. Therefore, the SU(2) group acts only on the left component of the fermion fields and is also denoted as $SU(2)_L$. The interaction term that couples left-chirality fermions to the gauge fields, obtained on expanding Eq. 2.14, is given by $-g_w\bar{\Psi}_L\gamma^\mu T_i \Psi_L W^i_\mu$ and is proportional to the coupling constant, g_w.

2.2.5 Electroweak Unification

While the boson fields, $W^\pm_\mu$, corresponding to the physical $W^\pm$ bosons, can be recovered from the W^i_μ boson fields in Eq. 2.14 using the linear combinations

$$W^\pm_\mu = \frac{1}{\sqrt{2}}\left(W^1_\mu \mp i W^2_\mu\right), \tag{2.17}$$

an additional U(1) group needs to be introduced to recover the Z boson field, Z_μ, from the W^3_μ field. This is done by introducing a gauge field B_μ under U(1) symmetry, similar to the one for QED. The quantum number associated to this field is called the weak hypercharge and is denoted by Y, and hence the symmetry is denoted as

$U(1)_Y$. The group has a single generator $Y/2$ and its Lagrangian is similar to that of QED and is given by

$$\mathcal{L}_Y = i\bar{\psi}\gamma^\mu D_\mu\psi - \frac{1}{4}B_{\mu\nu}B^{\mu\nu}, \tag{2.18}$$

omitting the mass term, where the covariant derivative is given by

$$D_\mu = \partial_\mu + ig_Y\frac{Y}{2}B_\mu, \tag{2.19}$$

with g_Y as the coupling constant corresponding to the weak hypercharge. The tensor field is defined in a similar fashion as in QED,

$$B_{\mu\nu} = \partial_\mu B_\nu - \partial_\nu B_\mu. \tag{2.20}$$

It can be noted from Eq. 2.18 that unlike the $SU(2)_L$ group, the $U(1)_Y$ group acts on the fermion field ψ of either chirality, albeit with a strength that depends on the hypercharge of the fermion. Therefore, the $U(1)_Y$ part of the weak interaction is not restricted exclusively to fermions with one chirality. The interaction term on expanding Eq. 2.18 takes the form $-g_Y\bar{\psi}\gamma^\mu(Y/2)\psi B_\mu$.

Combining the Lagrangians in Eqs. 2.14 and 2.18 one can define a Lagrangian for a combined $SU(2)_L \times U(1)_Y$ symmetry group, as

$$\mathcal{L}_{\text{EWK}} = \bar{\Psi}_L\gamma^\mu D_\mu\Psi_L + i\bar{\psi}_R\gamma^\mu D_\mu\psi_R + i\bar{\psi}'_R\gamma^\mu D_\mu\psi'_R - \frac{1}{4}W^i_{\mu\nu}W_i^{\mu\nu} - \frac{1}{4}B_{\mu\nu}B^{\mu\nu}, \tag{2.21}$$

where the covariant derivative now takes the form[1]

$$D_\mu = \partial_\mu + ig_w T^i_\mu W^i_\mu + ig_Y\frac{Y}{2}B_\mu. \tag{2.22}$$

The four bosons, W^i_μ and B_μ, corresponding to the $SU(2)_L \times U(1)_Y$ group can now be suitably "mixed" to express the physical boson fields as linear combinations of the former. While the $W^\pm_\mu$ fields are given by Eq. 2.17, a linear combination of the W^3_μ and B_μ fields gives rise to the physical Z_μ boson field. The remaining component of the mixing can be interpreted as the electromagnetic photon field, A_μ. This can be expressed as a rotation by angle θ_W, as:

$$\begin{pmatrix} A_\mu \\ Z_\mu \end{pmatrix} = \begin{pmatrix} \cos\theta_W & \sin\theta_W \\ -\sin\theta_W & \cos\theta_W \end{pmatrix} \begin{pmatrix} B_\mu \\ W^3_\mu \end{pmatrix}, \tag{2.23}$$

[1] The Lagrangians in Eqs. 2.14 and 2.18 were constructed independently, each containing terms of the form $i\bar{\psi}\gamma^\mu\partial_\mu\psi$, and hence naively adding them results in the terms of the form $i\bar{\psi}\gamma^\mu\partial_\mu\psi$ appearing twice. When combining, it is to be noted that the partial derivative in the *combined* Lagrangian is to be replaced with the covariant derivative containing multiple vector boson fields (e.g. in Eq. 2.22). Thus, each term of the form $i\bar{\psi}\gamma^\mu\partial_\mu\psi$ appears only once in the combined Lagrangian.

where θ_W is known as the Weinberg angle. The quantum numbers of the QED and weak interactions, namely, Q, I_3, and Y, are related by the so-called Gell-Mann-Nishijima formula [22, 23],

$$Q = I_3 + \frac{Y}{2}. \tag{2.24}$$

The coupling constants, g_w and g_Y, and the electric charge, e, are also related as

$$e = g_w \cos\theta_W = g_Y \sin\theta_W. \tag{2.25}$$

Thus, it is possible to recover the electromagnetic interaction from the electroweak unification, making the standalone QED approach described in Sect. 2.2.3 redundant.

The various terms obtained on expanding the electroweak Lagrangian in Eq. 2.21 can be grouped as follows:

- Kinetic terms of the form $\bar{\psi}\partial_\mu\psi$ that describe free-field kinematics.
- Terms involving the W^1_μ and W^2_μ boson fields (or equivalently, $W^\pm$ bosons) represent so-called charged current (CC) interactions. These couplings involve only left-chiral fermion fields, and hence parity is maximally violated.
- Terms involving the W^3_μ and B_μ boson fields represent so-called neutral current (NC) interactions. The W^3_μ coupling terms involve only left-chiral fermion fields, while the B_μ terms involve both left- and right-chiral ones. However, using Eq. 2.23, one can rearrange these terms and identify separate terms associated with the photon and Z boson fields. These interactions can involve both left- and right-chiral fermion fields.
- The kinetic terms corresponding to the propagation of the boson fields show a large number of self-interactions of the four gauge bosons. These can be trilinear (ZW^+W^-, γW^+W^-) or quadrilinear (ZZW^+W^-, $\gamma\gamma W^+W^-$, γZW^+W^-, $W^+W^-W^+W^-$).

It can be noted that the Lagrangian in Eq. 2.21 does not contain any mass terms. This is a necessity at this stage, as mass terms would violate the gauge invariance. The apparent conflict with the experimental observations of massive fermions, and Z and W bosons is resolved with the Higgs mechanism, as explained in Sect. 2.3.1.

Interactions mediated by the W boson can change the flavour of an up-type quark to a down-type one, making CC interaction the only interaction that can change flavours. On the other hand, the Z boson, although electrically neutral itself, couples to all charged fermions of any chirality under weak interactions. However, such interactions cannot change flavours of quarks.

2.2.6 *Phenomenological Consequences*

This subsection briefly discusses the effects of the discussions in the previous subsections on hadron collider experiments.

Strong interaction Only colour-charged particles participate in the strong interaction. The generators of the QCD group do not commute. Unlike the photon being neutral when it comes to electrical charges, gluons themselves contain colour charges. Based on colour combinations, there can be 8 colours of gluons. Gluons can also participate in self-interactions. Again in contrast to QED, where the electrical field between two electrically charged particles decreases with distance, the gluon field between a pair of colour charges becomes stronger as the distance between them increases. Hence, quarks always exist in colour-neutral bound states, called hadrons. If external energy is provided to separate the quarks in a bound state, after a certain energy threshold it becomes energetically favourable for the hadron to create a new $q\bar{q}$ pair and therefore a new hadron. This phenomenon is referred to as QCD colour confinement.

Colour confinement has far reaching consequences at hadron colliders. Any quarks or gluons generated at the hard collision immediately hadronise to form a chain of hadrons. Such quarks generated at high energies result in a collimated stream of hadrons, known as jets, being produced in the final state.

Electromagnetic interaction All electrically charged particles in the SM can interact with the QED field. As explained previously, electric charge remains conserved under QED interactions. The photon itself is a massless and chargeless particle. At the LHC, QED interactions are relevant in the context of detecting charged particles when they interact electromagnetically with detector material. Photons can also be detected in the same way due to electron-positron pairs being produced from photons interacting with detector material.

Weak interaction Neutrinos are neither electrically charged nor colour-charged. Therefore, neutrinos (specifically the left-chirality ones) can interact with detector material only through the weak interaction. However, such neutrino-nucleon interactions are extremely improbable and they are considered practically undetectable in conventional hadron collider experiments. On the other hand, right-chirality neutrinos, if they exist, are not subject to any interaction in the SM. They are considered as candidates for dark matter in some BSM theories [24].

2.3 The Higgs Boson

The Higgs mechanism involves the introduction of a complex scalar field, which along with a spontaneous symmetry breaking mechanism, imparts masses to the W and Z bosons. This Higgs field being an additional quantum field gives rise to an additional boson, called the Higgs boson, which has a spin 0. The following subsection describes the Higgs mechanism in more details, followed by subsections describing the phenomenology of the Higgs boson in the context of the LHC.

2.3.1 The Higgs Mechanism

The descriptions of the SM discussed in Sect. 2.2 do not explain the masses of the fermion and boson fields, as introduction of such mass terms would violate the gauge invariance. The Brout-Englert-Higgs (BEH) mechanism [25–27], also known as the electroweak symmetry breaking (EWSB) or the Higgs mechanism, was introduced as a mechanism to generate the gauge boson masses in the SM. The mechanism introduces two complex scalar fields (ϕ^+, ϕ^0) grouped as a weak isospin doublet called the Higgs field. The Higgs field is given by

$$\phi = \begin{pmatrix} \phi^+ \\ \phi^0 \end{pmatrix} = \frac{1}{\sqrt{2}} \begin{pmatrix} \phi_1 + i\phi_2 \\ \phi_3 + i\phi_4 \end{pmatrix}, \tag{2.26}$$

with ϕ_i as four real scalar fields. The SM Lagrangian is then supplemented with a Higgs Lagrangian of the form

$$\mathcal{L}_{\text{Higgs}} = \left(D_\mu\phi\right)^\dagger \left(D_\mu\phi\right) - V(\phi), \tag{2.27}$$

where D_μ is the covariant derivative introduced in Eq. 2.22 and $V(\phi)$ is a scalar potential given by

$$V(\phi) = -\mu^2\phi^\dagger\phi + \lambda\left(\phi^\dagger\phi\right)^2, \tag{2.28}$$

where μ^2 and λ are positive constants. This potential has a "Mexican-hat" shape, as shown in Fig. 2.2, such that at $\phi = 0$, it presents an unstable local maximum. On the other hand, the ground state with minimum energy is given by a continuum of doublets that minimise the potential at

$$\left|\phi^\dagger\phi\right| = \frac{\mu^2}{2\lambda} = \frac{v^2}{2}, \tag{2.29}$$

where v is known as the vacuum expectation value (VEV) of the field. As the system "chooses" a ground state among all possible energy-minimum states to achieve stability, it is said to spontaneously break the symmetry, while the Lagrangian remains gauge invariant. This choice of a ground state explicitly breaks both the $SU(2)_L$ and the $U(1)_Y$ symmetries, but preserves the $U(1)_{\text{QED}}$ symmetry when the ground state is expressed in a way such that it is parallel to the ϕ^0 component of the doublet, as

$$\phi_{\text{ground}} = \frac{1}{\sqrt{2}} \begin{pmatrix} 0 \\ v \end{pmatrix}. \tag{2.30}$$

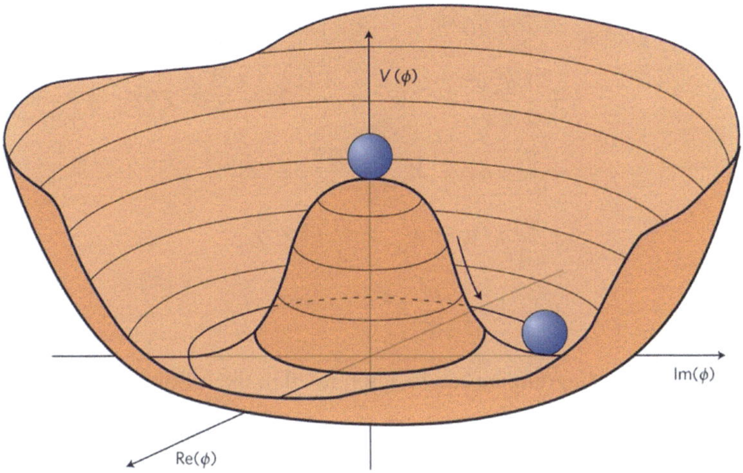

Fig. 2.2 The Higgs potential, with a non-zero unstable local maximum at the origin, and a stable state (minimum) away from the origin. Image by CERN [28]

The Higgs field can be perturbed around the minimum as

$$\phi_{\text{ground}} \rightarrow \phi(x) = \frac{1}{\sqrt{2}} e^{i\sigma_a \theta^a(x)} \begin{pmatrix} 0 \\ v + h(x) \end{pmatrix}, \tag{2.31}$$

where $\theta^a(x)$ $(a = 1, 2, 3)$ are three massless scalar fields, predicted by the Goldstone theorem [29–32], and $h(x)$ is a massive scalar field. The field $h(x)$ can be associated with a massive scalar boson, called the Higgs boson. However, the massless fields $\theta^a(x)$, associated to the so-called Goldstone bosons, are unphysical. They can be absorbed by means of an $SU(2)_L$ local transformation, under which the Lagrangian is invariant. This transformation of the Higgs field is referred to as the unitary gauge and can be represented as

$$\phi(x) \rightarrow e^{-i\sigma_a \theta^a(x)} \phi(x) = \phi'(x) = \frac{1}{\sqrt{2}} \begin{pmatrix} 0 \\ v + h(x) \end{pmatrix}, \tag{2.32}$$

such that $\phi(x)$ now depends only on the VEV and $h(x)$. The three Goldstone bosons are now transformed to reflect the longitudinal components of the physical $W^{\pm}$ and Z bosons.

Using Eq. 2.32 and adding kinetic terms for the $W^{\pm}$ and Z fields, the Lagrangian in Eq. 2.27 can now be rewritten as

$$\begin{aligned}
\mathcal{L}_{\text{Higgs}} = & \frac{1}{2}\partial_\mu h \partial^\mu h - \frac{1}{2}\left(2\mu^2\right) h^2 \\
& + \frac{1}{2}\frac{g_w^2 v^2}{2} W^{\mu+} W_\mu^- + \frac{1}{2}\frac{\left(g_w^2 + g_Y^2\right) v^2}{4} Z^\mu Z_\mu \\
& + \frac{g_w^2 v}{2} h W^{\mu+} W_\mu^- + \frac{g_Y^2 v}{2} h Z^\mu Z_\mu \\
& + \frac{g_w^2}{4} h^2 W^{\mu+} W_\mu^- + \frac{g_Y^2}{4} h^2 Z^\mu Z_\mu \\
& + \frac{\mu^2}{v} h^3 + \frac{\mu^2}{4v^2} h^4.
\end{aligned} \tag{2.33}$$

The terms in this Lagrangian can be interpreted in the following way:

- The first line represents the evolution of a scalar field, namely the Higgs field. The associated squared mass term is given by

$$m_{\text{H}}^2 = 2\mu^2 = 2\lambda v^2, \tag{2.34}$$

  and is a free parameter of the SM.
- The second line gives the mass terms for the $\text{W}^\pm$ and Z bosons:

$$\begin{aligned}
m_{\text{W}} &= \frac{g_w v}{2} \\
m_{\text{Z}} &= \frac{\sqrt{g_w^2 + g_Y^2}}{2} v = \frac{m_{\text{W}}}{\cos\theta_W}.
\end{aligned} \tag{2.35}$$

- The third line describes the trilinear interactions, λ_{HVV}, between the Higgs fields and the weak bosons:

$$\lambda_{\text{HVV}} = \frac{2}{v} m_{\text{V}}^2, \tag{2.36}$$

  where m_{V} represents mass of the boson, V ($\text{V}=\text{W}^\pm$, Z). These include interactions of the form HW^+W^- and HZZ.
- The fourth line describes the quadrilinear interactions, λ_{HHVV}, between the Higgs fields and the weak bosons:

$$\lambda_{\text{HHVV}} = \frac{1}{v^2} m_{\text{V}}^2. \tag{2.37}$$

  These include interactions of the form HHW^+W^- and HHZZ.
- The fifth (last) line describes the trilinear ($\lambda_{\textbf{HHH}}$) and quadrilinear (λ_{HHHH}) self-interactions of the Higgs boson:

$$\lambda_{\text{HHH}} = \lambda_{\text{HHHH}} = \frac{m_{\text{H}}^2}{v^2}. \tag{2.38}$$

The scalar potential in Eq. 2.28 can, therefore, be rewritten as

$$\begin{aligned} V(h) &= \frac{1}{2}\left(2\mu^2\right)h^2 + \frac{\mu^2}{v}h^3 + \frac{\mu^2}{4v^2}h^4 \\ &= \frac{1}{2}m_{\mathrm{H}}^2 h^2 + \lambda_{\mathrm{HHH}} v h^3 + \lambda_{\mathrm{HHHH}} h^4. \end{aligned} \tag{2.39}$$

Yukawa interaction The Higgs mechanism described above explains how $W^{\pm}$ and Z bosons acquire their masses. The masses of fermions can also be explained using the Higgs mechanism by using a so-called Yukawa interaction that couples a fermion field to the scalar Higgs fields. The coupling of the left-chirality doublet (Ψ_L) and two right-chirality singlet (ψ_R, ψ'_R) fields to the Higgs field (ϕ) can be incorporated using the Lagrangian

$$\mathcal{L}_{\text{Yukawa}} = -y_{f'}\left(\bar{\Psi}_L \phi \psi'_R + \bar{\psi}'_R \phi^{\dagger} \Psi_L\right) - y_f\left(\bar{\Psi}_L \phi^c \psi_R + \bar{\psi}_R \phi^{c\dagger} \Psi_L\right), \tag{2.40}$$

where y_f and $y_{f'}$ are the Yukawa coupling constants for up-type ($I_3 = +1/2$) and down-type ($I_3 = -1/2$) fermions, respectively, and $\phi^c = i\sigma_2\phi^*$ represents the charge conjugate of the Higgs field. After electroweak symmetry breaking one gets:

$$\mathcal{L}_{\text{Yukawa}} = -\sum_f m_f \bar{\psi}\psi\left(1 + \frac{h}{v}\right) - \sum_{f'} m_{f'} \bar{\psi}'\psi'\left(1 + \frac{h}{v}\right),$$

with the sums running separately over up- and down-type fermions. The first term in each sum represents the mass of the fermion,

$$m_{f^{(\prime)}} = y_{f^{(\prime)}} \frac{v}{\sqrt{2}}, \tag{2.41}$$

and the second represents coupling of the fermion to the Higgs field. Equation 2.41 shows that the mass of a fermion is proportional to the strength of the Yukawa coupling ($y_{f^{(\prime)}}$) of the fermion, and the VEV value, v. These masses are free parameters of the SM and can be measured experimentally, whence the value of $y_{f^{(\prime)}}$ can be inferred. As right-chirality neutrinos do not exist in the SM, this mechanism does not generate neutrino masses and hence neutrinos are massless in the SM. However, neutrino masses can be described by introducing right-chirality neutrino fields to the SM, so that neutrinos acquire masses in the same way as other fermions do.

Summary The Higgs mechanism introduces a scalar Higgs field that has a non-zero VEV. The interaction of gauge bosons and fermions with this field generates their masses without violating gauge invariance. The Higgs boson also self-interacts through the Higgs potential. Thus, the Higgs boson couples to all massive particles in the SM. While it cannot couple directly to massless photons, it can indirectly couple to them via loop interactions, involving, for instance, massive W bosons or top quarks in the loop.

The mass of the Higgs boson is a free parameter in the SM and has been experimentally measured to be 125.38 ± 0.14 GeV [33]. The several production and decay channels of the Higgs boson are listed in the following subsections.

2.3.2 *Higgs Boson Production at the LHC*

At pp collisions at the LHC at a centre-of-mass of several TeVs, the major production processes of the Higgs boson are as follows:

1. **Gluon fusion**: The dominant production mechanism involves the fusion of two gluons and a virtual loop of quarks, mainly consisting of the heavier top and bottom quarks. This process is denoted as ggH and has a production cross section of 48.6 pb at $\sqrt{s} = 13$ TeV. The process is depicted in Fig. 2.3a.
2. **Vector boson fusion**: Two Z bosons or two oppositely-charged W bosons radiated from two quarks in the initial state "fuse" to produce a Higgs boson. The final state additionally contains two quarks. This process is denoted as VBF Higgs, or simply VBF, and has a production cross section of 3.78 pb. It is depicted in Fig. 2.3b.
3. **Vector boson associated production**: The third dominant process involves a Higgs boson radiating off a Z or W boson (referred to as a Vector (V) boson together), which is, in turn, produced from a quark-antiquark pair annihilation in the hard collision. This is known as Higgs-strahlung and is depicted in Fig. 2.3c. The ZH final state can also be obtained from a top-quark loop induced by the interaction of pair of gluons, as shown in Fig. 2.3d. In either case, the final state consists of the Higgs boson alongside a leptonically- or hadronically-decaying

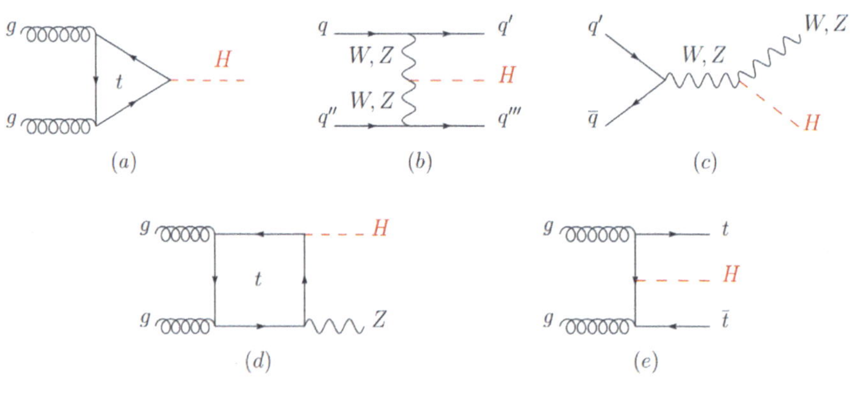

Fig. 2.3 Main leading order Feynman diagrams contributing to the Higgs boson production in **a** gluon fusion, **b** Vector boson fusion, **c** Higgs-strahlung (or associated production with a gauge boson at tree level from a quark-quark interaction), **d** associated production with a gauge boson (at loop level from a gluon-gluon interaction), **e** associated production with a pair of top quarks. Image from 2022 Review of Particle Physics by Particle Data Group [34, 35]

vector boson. This production mode is denoted as VH production and is the main production mode described in this thesis. The cross section of this process is 1.37 pb in the W-associated production mode and 0.88 pb in the Z-associated mode.

4. **Top pair associated production**: Two gluons in the initial state each produce a top quark-antiquark pair, whence a top quark and a top antiquark from the two pairs fuse to produce a Higgs boson. The final state comprises a Higgs boson and a top quark-antiquark pair. This process is usually denoted as ttH and has a production cross section of 0.50 pb. It is depicted in Fig. 2.3e.

While the ggH process has a production cross section that is an order of magnitude higher than that of the VH process, the former is more difficult to identify in hadronic Higgs boson decays due to an overwhelming background of multijet production processes (so-called QCD processes). The latter, despite having a smaller cross section, is easier to detect at hadron colliders. This is because leptonic decays of the vector boson in VH events can be "triggered" at hadron colliders using charged lepton or missing energy triggers (cf. Sect. 3.2.6). Moreover, leptonic-V final states of the VH process have smaller backgrounds compared to all-hadronic final states like ggH and VBF.

2.3.3 Higgs Boson Decays

The prominent decay modes of the Higgs boson are listed as follows:

- **Bottom quarks**: The $H \rightarrow b\bar{b}$ process is the most common decay of the Higgs boson with a branching ratio (BR) of 58.2%. However, such decays are rather hard to reconstruct due to limited efficiency of bottom quark identification. Moreover, overwhelming QCD backgrounds in the ggH and VBF production modes also make it difficult to measure the $H \rightarrow b\bar{b}$ process. The process was observed in 2018 [36, 37], making it one of the most recent observations of a Higgs boson decay mode.
- **W bosons**: The Higgs boson decays to a pair of W bosons with a BR of 21.4%, where at least one of the W bosons is off-shell. Such decays are usually studied in events containing leptonic decays of the W boson(s) [38–40].
- **Gluons**: The Higgs boson decays into a pair of gluons with a BR of 8.2%. But such decays are inaccessible at hadron colliders as separation of gluons final-states from QCD backgrounds is practically infeasible.
- **Taus**: The Higgs boson decays to a pair of tau leptons with a BR of 6.2% [41–43].
- **Charm quarks**: The $H \rightarrow c\bar{c}$ process has a BR of 2.9%. This BR is about $\sim$20 times smaller than that of the $H \rightarrow b\bar{b}$ decay channel. A low BR, coupled with the fact that charm quarks are even more challenging to identify than bottom quarks, makes this channel very challenging to probe. This thesis focuses on this decay channel of the Higgs boson.

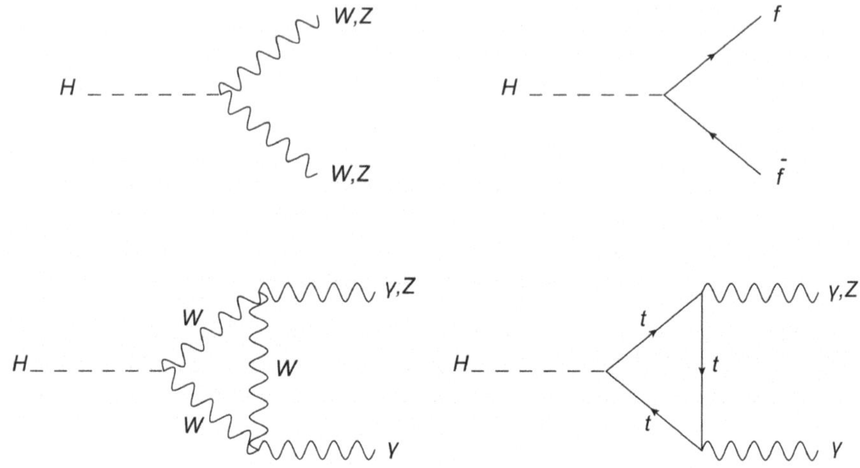

Fig. 2.4 Feynman diagrams for the Higgs boson decays into heavy vector boson pairs (upper-left), fermion–antifermion pairs (upper-right), and photon pairs and Zγ (lower)

- **Z bosons**: The decay to a pair of Z bosons has a BR of 2.6%, where at least one of the Z bosons is off-shell. Despite that, the four-lepton final state, where both Z bosons decay into charged leptons, is one of the cleanest and easiest channels to observe the Higgs boson [44, 45].
- **Photons**: The H $\rightarrow \gamma\gamma$ decay has a BR of only 0.2%. However, the excellent experimental resolution of the photon momenta makes this an important channel to study the Higgs boson [46, 47].
- **Z+photon**: The H $\rightarrow$ Zγ decay has a BR of only 0.15% and can be reconstructed from leptonic decay modes of the Z boson [48].
- **Muons**: The H $\rightarrow \mu^+\mu^-$ decay has a BR of 2×10^{-4}. While detecting a decay with such a small BR is challenging, the channel benefits from clean final states and excellent muon reconstruction capabilities [49].
- **Other decay modes**: Decays of the Higgs boson to electrons and lighter quarks are insignificant.

Figure 2.4 shows the Feynman diagrams for the decays of the Higgs boson to W and Z bosons, decay to fermions, and decays to $\gamma\gamma$ and Zγ.

2.4 Diboson Processes

The diboson process refers to the production of two vector bosons in a collision event, that is, a WW, ZZ, or WZ final state. The principal production mode of the processes is through the so-called s-channel, involving a quark-antiquark pair in the initial state annihilating to produce a vector boson, which decays to a pair of vector bosons in

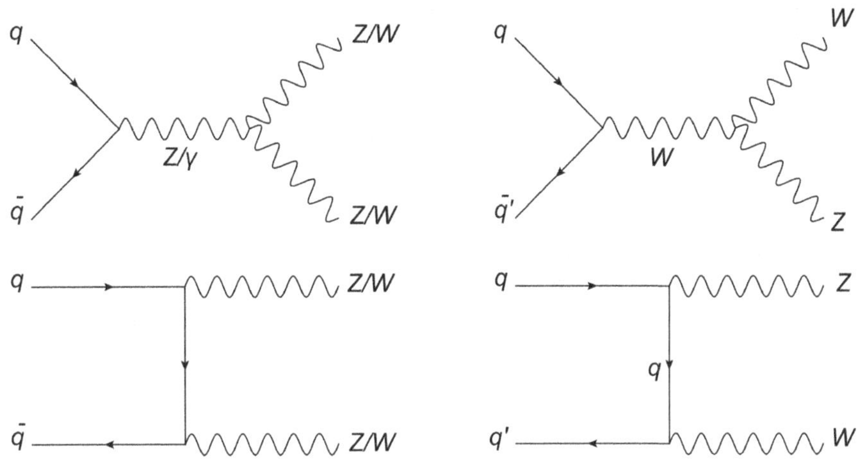

Fig. 2.5 Representative Feynman diagrams for ZZ and WW (left), and WZ (right) processes. The upper and lower rows represent the s- and t-channels, respectively

the final state. This results in a three-boson vertex. This channel can be compared to the VH production mode by replacing one of the final state vector bosons with a Higgs boson.

The t- and u-channel productions of this process are subdominant contributions. Gluon-gluon initial states also produce diboson events but have an insignificant contribution. The measured cross sections of the ZZ, WZ, and WW processes at $\sqrt{s} = 13$ TeV are 17.3 pb [50], 51.0 pb [51], and 142 pb [52], respectively. Figure 2.5 shows representative Feynman diagrams for diboson processes.

The Wγ and Zγ processes are also diboson processes, but have not been considered in the presented analyses.

2.5 The Background Processes

The major background processes relevant to the presented analyses are listed as follows:

- **Drell–Yan + jets**: The production of an opposite-sign charged-lepton pair in a hadron collision event is referred to as the Drell–Yan (DY) process [53]. The process is generally initiated by a quark-antiquark pair annihilating into a photon or a Z boson, which then pair-produces charged leptons. Such processes are often associated with one or more jets arising from initial-state gluon or quark radiations (ISRs), or from proton remnants. Alternatively, a quark-gluon interaction can also produce a Z+jet event. Such DY+jets events form an important background in the presented analyses.

- **Leptonic W + jets**: Processes with a leptonically-decaying W boson produced in association with one or more jets also form an important background in the presented analyses. Such a process is usually initiated by a gluon and a quark, or a pair of quarks interacting in the initial state.
- **Z($\nu\nu$) + jets**: The production of a Z boson decaying into neutrinos in association with jets forms an important background in the presented analyses. The production mechanism of this process is very similar to that of the Z-mediated DY process, while the decay products of the Z boson are different.
- **Top quark-antiquark pair**: The production of a top quark-antiquark pair is referred to as the $t\bar{t}$ process. The dominant production mode of the $t\bar{t}$ process at the LHC is from the interaction of a pair of gluons, but it is also produced from an initial-state quark-antiquark pair annihilating into a gluon, that further splits into a top quark-antiquark pair. Each top quark immediately decays into a lighter quark (almost always a bottom quark) plus a W boson. Each W boson can decay either hadronically or leptonically. Therefore, the $t\bar{t}$ process can be either fully hadronic, semileptonic, or fully leptonic (dileptonic). The measured inclusive cross section is 746 pb at $\sqrt{s} = 13$ TeV [54].
- **Single top**: The single top process refers to the production of a single top quark in association with an additional quark or a W boson. It forms a minor background in a number of channels used in the presented analyses. The tW process has a measured cross section of 89 pb [55], while the t-channel and s-channel single top processes have cross sections of 238 pb [56] and 14.6 pb [57], respectively, at $\sqrt{s} = 13$ TeV.
- **Diboson**: Diboson processes, as described in Sect. 2.4, form a minor background in the Higgs boson search presented in this thesis.

Representative Feynman diagrams for the V+jets processes are shown in Fig. 2.6. The theoretical cross sections of these processes with various final states are listed in Tables B.6 and B.7 in Appendix B. Feynman diagrams for $t\bar{t}$ and single top are shown in Fig. 2.7.

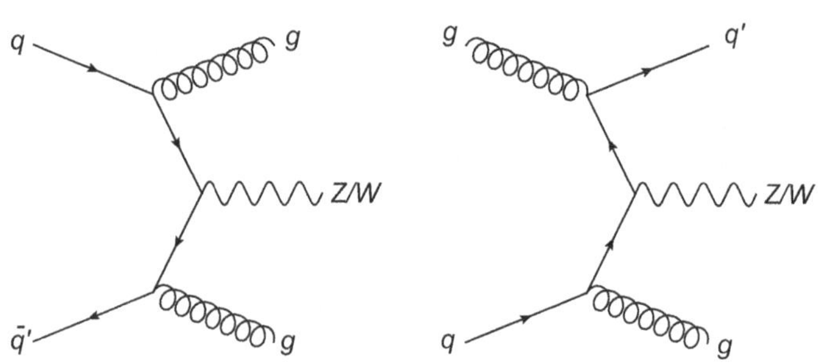

Fig. 2.6 Representative Feynman diagrams for vector boson production in association with 2 jets at the LHC

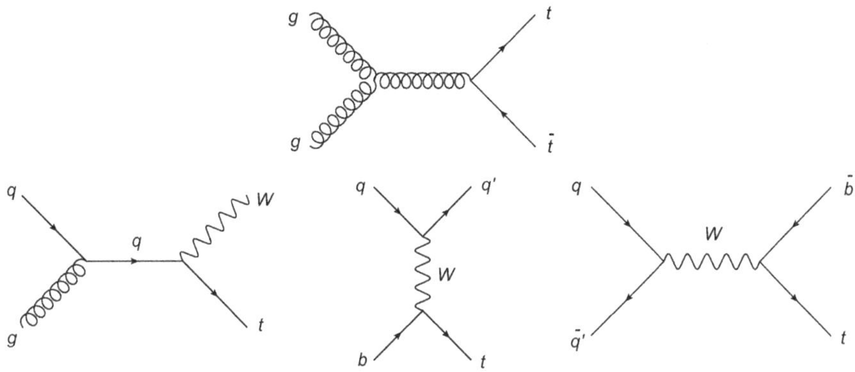

Fig. 2.7 Representative Feynman diagrams for top quark-antiquark pair production (upper) and single top production (lower) at the LHC. Three production modes of the single top process have been shown, namely tW process (left), t-channel (middle), and s-channel (right)

References

1. Griffiths D (1987) Introduction to elementary particles. Wiley
2. Halzen F, Martin AD (1984) Quark and leptons: an introductory course in modern particle physics. Wiley
3. Peskin ME, Schroeder DV (1995) An introduction to quantum field theory. Addison-Wesley
4. Weinberg S (1996) The quantum theory of fields. Cambridge University Press
5. 't Hooft G, Veltman MJG (1972) Regularization and renormalization of gauge fields. Nucl Phys B 44:189–213. https://doi.org/10.1016/0550-3213(72)90279-9
6. Poincaré MH Sur la dynamique de lélectron. Rendiconti del Circolo Matematico di Palermo (1884–1940) 21:129–175
7. Minkowski H (1910) Die grundgleichungen für die elektromagnetischen vorgänge in bewegten körpern. Mathematische Annalen 68:472–525. https://doi.org/10.1007/BF01455871
8. Fermi E On the quantization of the monoatomic ideal gas. arXiv:cond-mat/9912229. An English translation by A. Zannoni (Mayrland U.) of the original work of Enrico Fermi
9. Dirac P (1926) On the Theory of quantum mechanics. Proc R Soc Lond A 112:661–677. https://doi.org/10.1098/rspa.1926.0133
10. Bose SN (1924) Plancks gesetz und lichtquantenhypothese. Zeitschrift für Physik 26:178–181
11. Miss MJ, Cush (2019) Standard model of elementary particles. Wikimedia Commons. https://en.wikipedia.org/wiki/File:Standard_Model_of_Elementary_Particles.svg
12. Creative commons attribution 3.0 unported (CC BY 3.0). https://creativecommons.org/licenses/by/3.0/
13. LHCb Collaboration (2015) Observation of $J/\psi p$ resonances consistent with pentaquark states in $\Lambda_b^0 \to J/\psi K^- p$ decays. Phys Rev Lett 115:072001. https://doi.org/10.1103/PhysRevLett.115.072001. arXiv:1507.03414
14. LHCb Collaboration (2017) Observation of $J/\psi\phi$ structures consistent with exotic states from amplitude analysis of $B^+ \to J/\psi\phi K^+$ decays. Phys Rev Lett 118:022003. https://doi.org/10.1103/PhysRevLett.118.022003. arXiv:1606.07895
15. Davis R Jr, Harmer DS, Hoffman KC (1968) Search for neutrinos from the sun. Phys Rev Lett 20:1205–1209. https://doi.org/10.1103/PhysRevLett.20.1205
16. SNO Collaboration (2001) Measurement of the rate of $\nu_e + d \to p + p + e^-$ interactions produced by ^{8}B solar neutrinos at the sudbury neutrino observatory. Phys Rev Lett 87:071301. https://doi.org/10.1103/PhysRevLett.87.071301. arXiv:nucl-ex/0106015

17. Super-Kamiokande Collaboration (1998) Evidence for oscillation of atmospheric neutrinos. Phys Rev Lett 81:1562–1567. https://doi.org/10.1103/PhysRevLett.81.1562. arXiv:hep-ex/9807003
18. UA1 Collaboration (1983) Experimental observation of isolated large transverse energy electrons with associated missing energy at $\sqrt{s} = 540$ GeV. Phys Lett B 122:103–116. https://doi.org/10.1016/0370-2693(83)91177-2
19. UA2 Collaboration (1983) Observation of single isolated electrons of high transverse momentum in events with missing transverse energy at the CERN $\bar{p}p$ collider. Phys Lett B 122:476–485. https://doi.org/10.1016/0370-2693(83)91605-2
20. Weinberg S (1967) A model of leptons. Phys Rev Lett 19:1264–1266. https://doi.org/10.1103/PhysRevLett.19.1264
21. Wu CS et al (1957) Experimental test of parity conservation in beta decay. Phys Rev 105:1413–1415. https://doi.org/10.1103/PhysRev.105.1413
22. Nakano T, Nishijima K (1953) Charge independence for V-particles. Prog Theor Phys 10:581–582. https://doi.org/10.1143/PTP.10.581. arXiv:academic.oup.com/ptp/article-pdf/10/5/581/5364926/10-5-581.pdf
23. Gell-Mann M (1956) The interpretation of the new particles as displaced charge multiplets. Il Nuovo Cimento (1955–1965) 4:848–866. https://doi.org/10.1007/BF02748000
24. Drewes M (2013) The phenomenology of right handed neutrinos. Int J Mod Phys E Nucl Phys 22:1330019. https://doi.org/10.1142/s0218301313300191. arXiv:1303.6912
25. Englert F, Brout R (1964) Broken symmetry and the mass of gauge vector mesons. Phys Rev Lett 13:321. https://doi.org/10.1103/PhysRevLett.13.321
26. Higgs PW (1964) Broken symmetries, massless particles and gauge fields. Phys Lett 12:132–133. https://doi.org/10.1016/0031-9163(64)91136-9
27. Higgs PW (1964) Broken symmetries and the masses of gauge bosons. Phys Rev Lett 13:508–509. https://doi.org/10.1103/PhysRevLett.13.508
28. Creative commons attribution 4.0 international (CC BY 4.0). https://creativecommons.org/licenses/by/4.0/
29. Nambu Y (1960) Quasi-particles and gauge invariance in the theory of superconductivity. Phys Rev 117:648–663. https://doi.org/10.1103/PhysRev.117.648
30. Nambu Y, Jona-Lasinio G (1961) Dynamical model of elementary particles based on an analogy with superconductivity. I. Phys Rev 122:345–358. https://doi.org/10.1103/PhysRev.122.345
31. Goldstone J (1961) Field theories with Superconductor solutions. Il Nuovo Cimento (1955–1965) 19:154–164. https://doi.org/10.1007/BF02812722
32. Goldstone J, Salam A, Weinberg S (1962) Broken symmetries. Phys Rev 127:965–970. https://doi.org/10.1103/PhysRev.127.965
33. CMS Collaboration (2020) A measurement of the Higgs boson mass in the diphoton decay channel. Phys Lett B 805:135425. https://doi.org/10.1016/j.physletb.2020.135425. arXiv:2002.06398
34. Particle Data Group Collaboration (2022) Review of particle physics. PTEP 2022:083C01. https://doi.org/10.1093/ptep/ptac097
35. Attribution-noncommercial 4.0 international (CC BY-NC 4.0). https://creativecommons.org/licenses/by-nc/4.0/
36. CMS Collaboration (2018) Observation of Higgs boson decay to bottom quarks. Phys Rev Lett 121:121801. https://doi.org/10.1103/PhysRevLett.121.121801. arXiv:1808.08242
37. ATLAS Collaboration (2018) Observation of $H \to b\bar{b}$ decays and VH production with the ATLAS detector. Phys Lett B 786:59. https://doi.org/10.1016/j.physletb.2018.09.013. arXiv:1808.08238
38. CMS Collaboration (2014) Measurement of Higgs boson production and properties in the WW decay channel with leptonic final states. JHEP 01:096. https://doi.org/10.1007/JHEP01(2014)096. arXiv:1312.1129
39. ATLAS Collaboration (2015) Observation and measurement of Higgs boson decays to WW* with the ATLAS detector. Phys Rev D 92:012006. https://doi.org/10.1103/PhysRevD.92.012006. arXiv:1412.2641

40. ATLAS Collaboration (2015) Study of (W/Z)H production and Higgs boson couplings using H → WW* decays with the ATLAS detector. JHEP 08:137. https://doi.org/10.1007/JHEP08(2015)137. arXiv:1506.06641
41. CMS Collaboration (2014) Evidence for the 125 GeV Higgs boson decaying to a pair of τ leptons. JHEP 05:104. https://doi.org/10.1007/JHEP05(2014)104. arXiv:1401.5041
42. ATLAS Collaboration (2015) Evidence for the Higgs-boson Yukawa coupling to tau leptons with the ATLAS detector. JHEP 04:117. https://doi.org/10.1007/JHEP04(2015)117. arXiv:1501.04943
43. CMS Collaboration (2017) Observation of the Higgs boson decay to a pair of tau leptons. Phys Lett B 779:283. https://doi.org/10.1016/j.physletb.2018.02.004. arXiv:1708.00373
44. CMS Collaboration (2014) Measurement of the properties of a Higgs boson in the four-lepton final state. Phys Rev D 89:092007. https://doi.org/10.1103/PhysRevD.89.092007. arXiv:1312.5353
45. ATLAS Collaboration (2015) Measurements of Higgs boson production and couplings in the four-lepton channel in pp collisions at center-of-mass energies of 7 and 8 TeV with the ATLAS detector. Phys Rev D 91:012006. https://doi.org/10.1103/PhysRevD.91.012006. arXiv:1408.5191
46. CMS Collaboration (2014) Observation of the diphoton decay of the Higgs boson and measurement of its properties. Eur Phys J C 74:3076. https://doi.org/10.1140/epjc/s10052-014-3076-z. arXiv:1407.0558
47. ATLAS Collaboration (2014) Measurement of Higgs boson production in the diphoton decay channel in pp collisions at center-of-mass energies of 7 and 8 TeV with the ATLAS detector. Phys Rev D 90:112015. https://doi.org/10.1103/PhysRevD.90.112015. arXiv:1408.7084
48. CMS Collaboration Search for Higgs boson decays to a Z boson and a photon in proton-proton collisions at $\sqrt{s} = 13$ TeV. arXiv:2204.12945. Accepted for publication by JHEP
49. CMS Collaboration (2021) Evidence for Higgs boson decay to a pair of muons. JHEP 01:148. https://doi.org/10.1007/JHEP01(2021)148. arXiv:2009.04363
50. ATLAS Collaboration (2018) $ZZ \rightarrow \ell^+\ell^-\ell'^+\ell'^-$ cross-section measurements and search for anomalous triple gauge couplings in 13 TeV pp collisions with the ATLAS detector. Phys Rev D 97(3):032005. https://doi.org/10.1103/PhysRevD.97.032005. arXiv:1709.07703
51. ATLAS Collaboration (2019) Measurement of $W^{\pm}Z$ production cross sections and gauge boson polarisation in pp collisions at $\sqrt{s} = 13$ TeV with the ATLAS detector. Eur Phys J C 79(6):535. https://doi.org/10.1140/epjc/s10052-019-7027-6. arXiv:1902.05759
52. ATLAS Collaboration (2017) Measurement of the W^+W^- production cross section in pp collisions at a centre-of-mass energy of $\sqrt{s} = 13$ TeV with the ATLAS experiment. Phys Lett B 773:354–374. https://doi.org/10.1016/j.physletb.2017.08.047. arXiv:1702.04519
53. Drell SD, Yan T-M (1970) Massive lepton pair production in hadron-hadron collisions at high-energies. Phys Rev Lett 25:316–320. https://doi.org/10.1103/PhysRevLett.25.316. [Erratum: Phys Rev Lett 25:902 (1970)]
54. CMS Collaboration (2016) Measurement of the top quark pair production cross section in proton-proton collisions at $\sqrt{s} = 13$ TeV. Phys Rev Lett 116(5):052002. https://doi.org/10.1103/PhysRevLett.116.052002. arXiv:1510.05302
55. CMS Collaboration (2021) Observation of tW production in the single-lepton channel in pp collisions at $\sqrt{s} = 13$ TeV. JHEP 11:111. https://doi.org/10.1007/JHEP11(2021)111. arXiv:2109.01706
56. CMS Collaboration (2017) Cross section measurement of t-channel single top quark production in pp collisions at $\sqrt{s} = 13$ TeV. Phys Lett B 772:752–776. https://doi.org/10.1016/j.physletb.2017.07.047. arXiv:1610.00678
57. Muller D (2020) Search for single top quark production in the s channel at 13 TeV with the CMS experiment. Presented 05 Jun 2020. https://cds.cern.ch/record/2723842

Chapter 3
Experimental Setup

3.1 The Large Hadron Collider

The LHC [1–3] is the most powerful particle accelerator ever devised. Located in a circular underground tunnel of 26.7 km radius near Geneva at the Swiss-French border, this scientific marvel is designed to accelerate protons to energies up to 7 TeV, and lead atoms to up to 2.8 TeV per nucleon, which are the highest achievable energies till date. This subsection and rest of the thesis discuss the LHC in the context of pp collisions only.

The LHC was designed primarily to discover the Higgs boson. After 10 years of construction, the first particle collisions took place in 2009. In 2011 and 2012, the LHC operated at centre of mass energies of 7 and 8 TeV, respectively, and provided enough data for the discovery of the Higgs boson. After a period of repairs and upgrades in 2013 and 2014, the second run (Run-2) of the LHC began in 2015 at a centre of mass energy of 13 TeV. The LHC was again stopped for a period of 3 years between 2019 and 2022 for further upgrades before it resumed collisions for a third run.

Starting with hydrogen as a source of protons, a number of accelerators pre-accelerate protons before they are injected into the LHC for further acceleration. This series of accelerators include a linear accelerator (LINAC 2), the Proton Synchrotron Booster (PSB), the Proton Synchrotron (PS), and the Super Proton Synchrotron (SPS), which cumulatively accelerate protons to 450 GeV. Within the LHC, counter-rotating beams of protons are accelerated by resonant waves of electromagnetic fields, resulting in an energy gain of approximately 16 MeV per revolution per proton. Figure 3.1 shows the CERN accelerator complex.

The circular trajectory of the beams is maintained using superconducting dipole magnets that generate magnetic fields with strengths of up to 8.3 T. Additional multipole magnets are used to keep the beams focused despite electromagnetic repulsion between individual protons, such that the beams are confined to a diameter of the

S. Mondal, *Charming Decays of the Higgs, Z, and W Bosons*, Springer Theses,
https://doi.org/10.1007/978-3-031-84677-9_3

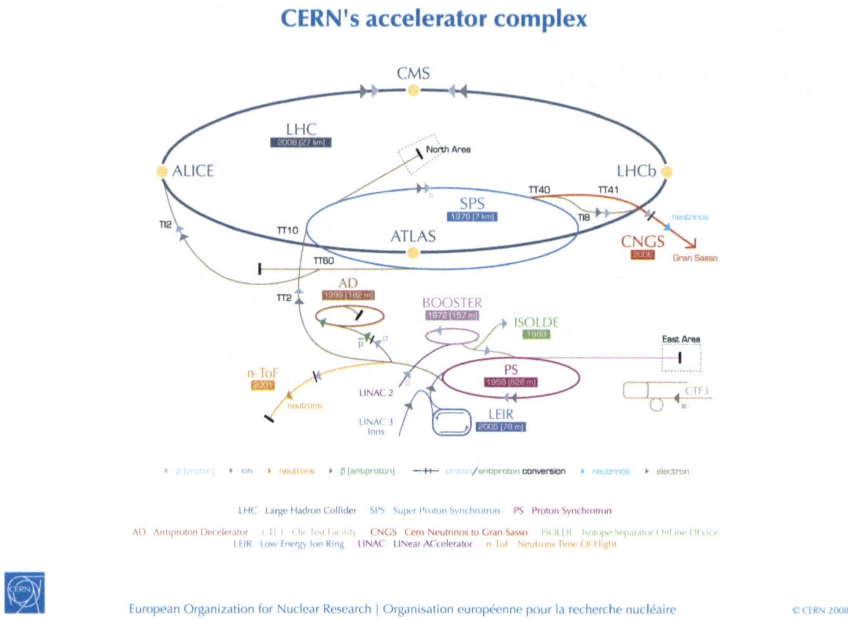

Fig. 3.1 The LHC is the last ring (dark grey line) in a complex chain of particle accelerators. The smaller machines are used in a chain to help boost the particles to their final energies and provide beams to a whole set of smaller experiments, which also aim to uncover the mysteries of the universe. Photograph and caption by Christiane Lefèvre, distributed by CERN

order of $\sim$10 μm at the collision points. The centre of mass energy of the collision of two identical, counter-rotating beams, each with energy E_{beam}, is given by

$$\sqrt{s} = 2 \cdot E_{\text{beam}}. \tag{3.1}$$

The other important parameter in collider physics is the instantaneous luminosity L, which is a function of several collider configuration parameters. It can be expressed as

$$L = \frac{N_b^2 n_b f \gamma F}{4\pi \epsilon_n \beta^*}, \tag{3.2}$$

where N_b is the number of particles per bunch, n_b is the number of bunches per beam, f is the revolution frequency of the beam, γ is the relativistic factor $\gamma = \frac{E}{m}$, F is the geometric reduction factor due to inclined beam profile, ϵ_n is the normalised transverse beam emittance, and β^*, the beta function at the interaction point, quantifies the transverse size of the particle beam. The LHC is capable of generating luminosities of upto 10^{34} cm^{-2}s^{-1}. The luminosity allows us to calculate the interaction rate $\frac{dN}{dt}$ of a process with cross section σ using the relation

$$\frac{dN}{dt} = \sigma \cdot L. \tag{3.3}$$

The integrated luminosity $\mathcal{L}$ is a measure of the total amount of data recorded and is defined as

$$\mathcal{L} = \int L dt, \tag{3.4}$$

which, along with Eq. 3.3, can be written as

$$N = \sigma \cdot \mathcal{L}. \tag{3.5}$$

There are a total of four interactions points along the circumference of the LHC where the trajectories of the proton bunches are slightly bent and made to collide. Around these interaction points, large particle detectors are placed to record the products of the collisions and store data for analysis. Two of these, the ATLAS [4] and CMS [5], are general-purpose detectors focusing on several aspects of physics. A third one, the ALICE detector [6] focuses on heavy lead ion collisions. The LHCb detector [7] specialises in analysing decays of long-lived b and c hadrons. In addition to these four, the TOTEM and LHCf detectors analyse forward collisions close to the beam line, near the CMS and ATLAS detectors, respectively. The MoEDAL experiment searches for hypothetical monopoles. Finally, FASER and SND are new detectors specialising in the search for new weakly interacting particles and detection of collider neutrinos, respectively.

3.2 The Compact Muon Solenoid Experiment

The CMS collaboration is one of the largest international collaborations, involving around 5500 physicists, engineers, technicians, students, and support staff from 241 institutes across 54 countries (as of May 2022). Its physics programme includes studying SM physics (including Higgs boson physics) as well as BSM physics, such as searching for extra dimensions and particles that could make up dark matter. It also studies collisions of heavy lead ions.

The CMS detector is a general-purpose detector at the LHC. It is located at one of the interaction points of the LHC, namely Point 5, at Cessy, France, approximately 100 m under the ground. It has a cylindrical shape and is around 21 m long along the beam axis and about 15 m in diameter. It accommodates several subdetectors aligned in layers around the beam axis, symmetric to the primary interaction point, and weighs around 14 thousand tonnes. As the name suggests, the detector is rather compact and utilises most of the available space to accommodate subdetectors. The central structure is a very strong solenoid magnet capable of generating a magnetic field with a strength of up to 4 T. An extensive system of muon chambers allows very precise detection of muons, justifying the detector's name.

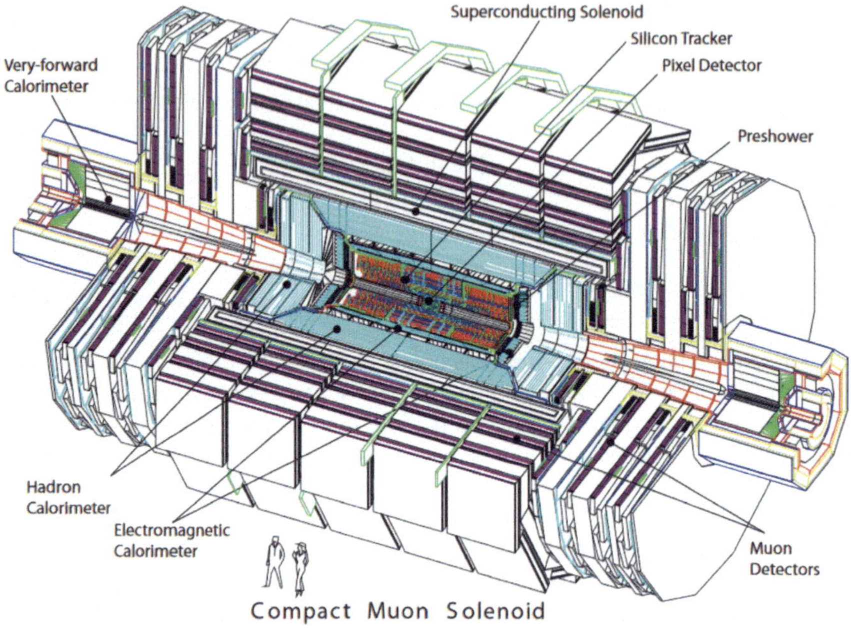

Fig. 3.2 A perspective view of the CMS detector [5]

Figure 3.2 shows an illustration of the CMS detector. A more detailed overview of the subdetectors is presented in the following subsections, followed by a description of the CMS trigger system.

3.2.1 Tracking System

The first layer, closest to the beam pipe, is the tracking system of the CMS detector. The primary purpose of this subdetector system is to reconstruct tracks of charged particles that are generated from pp collisions at the interaction point, utilising "hits" of the charged particles with the detector material [8]. As the tracker is located within the solenoid magnet, a homogeneous magnetic field across the entire tracker volume results in a curvature of the tracks of charged particles that move through the tracker. Hence, a precise reconstruction of tracks of such particles allows for a very precise measurement of their transverse momenta (p_T). Transverse impact parameter (IP) resolution of each track ranges from 20 to 75 mm depending on the p_T and pseudorapidity[1] (η) of the track. The general working principle is based on silicon semiconductors provided with a bias voltage such that transitting charged

[1] The pseudorapidity can be deduced from the polar angle, θ, using the relation $\eta = -\ln\left[\tan\left(\frac{\theta}{2}\right)\right]$, under the convention of the beam axis being along the z-axis. At high energies, η is equivalent to

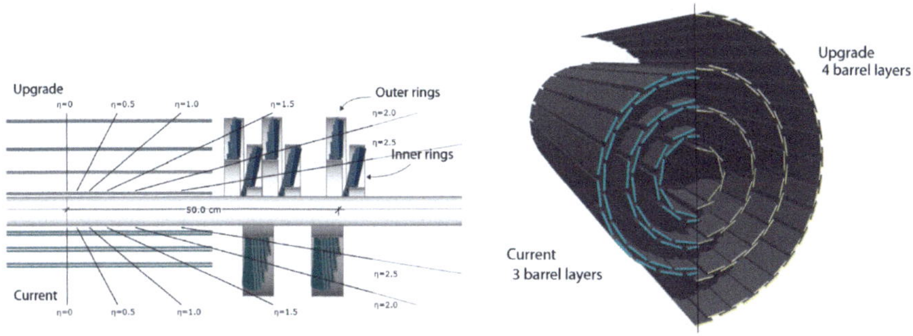

Fig. 3.3 Left: Layout comparing the different layers and disks in the Phase-0 and Phase-1 pixel detectors. Right: Transverse-oblique view comparing the pixel barrel layers in the two detectors [9]. "Current" in the diagram refers to the Phase-0 tracker, while "upgrade" refers to the Phase-1 tracker

particles induce electron-hole pairs. This, in turn, creates measurable currents which are subsequently digitised and electronically read out using cables. The resulting "hits" are grouped into tracks to reconstruct trajectories of each charged particle.

The tracking system consists of an inner pixel detector containing around 66 million pixels and an outer strip tracker containing around 9 million silicon microstrips with varying dimensions depending on their radial distance. The inner pixel detector consisted of three detector layers until the end of 2016 data-taking period (Phase-0) and was replaced with a four-layer detector before the 2017 data-taking period as a part of the Phase-1 detector upgrade [9]. Figure 3.3 provides an illustrative comparison of the trackers in the two phases. The upgraded high-efficiency and low-mass tracker provides four-hit pixel coverage to efficiently detect charged particles within the η range $|\eta| < 2.5$, while the Phase-0 inner tracker extended until $|\eta| < 2.4$. This upgrade led to a significant increase in track reconstruction efficiency and resolution in data collected in 2017 and 2018, compared to that in 2016 and earlier [10].

The tracking system plays a crucial role in the identification of jets initiated by c and b quarks. The identification algorithms explained in Sect. 6.2 rely on the efficient detection of tracks and their IPs. The CMS tracking system precisely reconstructs individual tracks from heavy hadron decays and therefore enables efficient bottom and charm tagging.

3.2.2 *Electromagnetic Calorimeter*

The electromagnetic calorimeter (ECAL) is installed around the tracking system. Its primary purpose is to absorb and measure the energies of electrons and photons

---

the rapidity, y, of a particle where $y = \frac{1}{2}\ln\left[\frac{E+p_z}{E-p_z}\right] = \text{arctanh}\left(\frac{p_z}{E}\right)$, with E and p_z as the particle's energy and z-component of the momentum, respectively. Differences in pseudorapidities ($\Delta\eta$) being a Lorentz-invariant quantity makes the use of η convenient in collider physics.

passing through it. The central part of the ECAL is a cylindrical barrel and covers the inner region until $|\eta| < 1.479$, while the outer part (endcap) covers the area between $1.479 < |\eta| < 3.0$. It consists of around 70 thousand scintillating lead tungstate crystals.

Both electrons and photons are measured in the ECAL. When these particles interact with the crystals, they create showers due to successive chains of EM radiation (photons), pair production (electrons and positrons), and annihilation. Charged particles passing through the material produce scintillation light, which is then amplified by photodiodes and phototriodes. The intensity of the amplified light gives a measurement of the energy of the inducing particle and is digitised and stored using read-out electronics.

3.2.3 Hadronic Calorimeter

The hadronic calorimeter (HCAL) is installed between the ECAL and the solenoid. However, to capture any part of hadronic activity that remains unabsorbed in the HCAL and the solenoid, a thin layer of HCAL outside the solenoid complements the bulk of the HCAL inside. The primary purpose of the HCAL is to absorb and measure hadrons (mesons and baryons) that originate as a result of hadronisation of quarks produced in the collision. It consists of alternating layers of non-magnetic brass absorbers and plastic scintillators.

When a hadron passes through the brass layers, it creates hadronic showers and induces scintillation light in the plastic scintillators. The light is then guided through optical fibres and read out through the electronics. Thus the HCAL plays an important role in detecting and measuring energies of hadronic particles that constitute jets.

3.2.4 Solenoid Magnet

The superconducting solenoid magnet is located radially outside the calorimeter system. It is 12.5 m in length and 6 m in diameter, making it the largest solenoid magnet ever constructed. The coil of the solenoid is made of superconducting NbTi and operates at a temperature of 4.7 K to maintain its superconductivity. It is designed to provide a constant magnetic field of 3.8 T uniformly in all subdetectors inside it. The magnetic field results in a curvature of trajectories of all electrically charged particles, thereby resulting in an excellent p_{T} resolution of charged particles in the tracker. In addition to the solenoid, three layers of massive steel flux-return yoke are installed to return the magnetic flux outside the solenoid back into the detector volume.

3.2.5 Muon Chambers

The muon system is the outermost layer of the CMS detector and is located radially outside the solenoid magnet. It consists of gas ionisation chambers embedded in the steel flux-return yoke outside the solenoid. When charged particles traverse the chambers, they produce electron avalanches that can be read out. Usually the only particles not absorbed by the inner layers of the detector are muons and neutrinos. As neutrinos cannot be detected, the hits in the muon chambers combined with the information from the tracker can be used to reconstruct muons.

The muon system uses three different types of detector approaches: Cathode Strip Chambers (CSCs), Drift Tubes (DTs), and Resistive Plate Chambers (RPCs). The CSCs are made up of anode wires and cathode strips placed perpendicularly. When a charged particle passes through the gas in this detector, the gas is ionised and a charge is induced on the cathode strips. The electrons from the induced charge move towards the anode wires creating an avalanche of ionisations. The gas in the DTs has a high drift velocity for ionised electrons, thereby enabling measurement of coordinates with a good spatial resolution by using the time difference between passage of particles and arrival of the electrons. In contrast, the RPCs provide a good time resolution. They are made with plates separated by a narrow area of gas and placed at a large voltage difference. Particles ionising the gas causes an immediate spark between the plates resulting in a very good time resolution. The excellent time resolution of the RPCs employed together with the good spatial resolution of the CSCs and DTs achieve an optimal resolution for both time and position measurements of muons.

3.2.6 Trigger System and Data Acquisition

The LHC at its peak luminosity and with branch crossings happening every 25 ns, performs almost 40 million pp collisions per second at the interaction point at CMS. The readouts from all the electronic channels across all subdetectors (event information) add up to a storage size of the order of $\sim$1 MB per event. This results in an output of the order of $\sim$40 TB per second when the LHC is operating at the expected luminosity. This data output rate far exceeds the processing and storing capabilities of even the most advanced data acquisition and storage systems ever developed. Moreover, a vast majority of events are low-energy soft collisions which are not interesting for the CMS physics programme. Therefore, it is necessary to store only a small fraction of the generated data that could potentially be interesting, for offline analysis. This task is achieved by employing a trigger system [11], which reads the detector outputs, makes a very fast decision based on a crude event reconstruction, and "triggers" the storage of only certain events that exhibit signatures of interesting physics.

The trigger system in CMS has two levels: the hardware-based Level 1 (L1) trigger system, and a software-based High-Level Trigger (HLT) system.

L1 trigger: Recorded data is held in the hardware buffers (pipeline) for about 3 μs. Immediately after a collision, signals from the readout electronics are sent to the L1 trigger processors. The L1 trigger [12] is implemented at a hardware level and makes decisions based on subdetector information. A muon trigger reads information from the muon chambers, while a calorimeter trigger reads information from the ECAL and the HCAL. The data processed at this level is "crude", that is, the data has lower granularity (lower resolution) than the capabilities of the subdetectors. The information processed by these are then combined to form a Global L1 trigger, which then makes a decision based on the reconstructed "trigger primitive objects", such as muons, electrons, photons, jets, and missing transverse energy. This decision is made in less than 1 μs and the result is then sent back to the pipeline, whence the data is either discarded or forwarded to the HLT for further filtering. The output rate of the L1 trigger is typically of the order of $\sim$100 kHz.

HLT: The data that passes the L1 trigger is read out and digitised by a dedicated switching system. The software-based HLT uses more sophisticated constraints on physics objects, which are reconstructed with higher precision combining the information from the tracking system. At the HLT level, several pre-defined "paths" are used to judge whether an event should be stored or discarded. Each of these paths typically combines a number of physics objects with certain kinematic and/or quality thresholds on each and is designed to specifically reflect a certain requirement of the CMS physics programme. An event is stored if it passes the requirements of at least one of the HLT paths, and discarded otherwise. The output rate of the HLT is of the order of $\sim$100 Hz.

Events that pass the trigger system are then stored offline and subsequently distributed across several computing nodes across the world for storage. The information of the L1 and HLT triggers passed is also stored for every event, and the events are usually grouped into non-exclusive sets of data (primary datasets) depending on which HLT path(s) the events have passed.

References

1. CERN LHC design report v.1: the LHC main ring. ebook: https://doi.org/10.5170/CERN-2004-003-V-1. http://cds.cern.ch/record/782076
2. CERN LHC design report v.2: the LHC infrastructure and general services. ebook: https://doi.org/10.5170/CERN-2004-003-V-2. http://cds.cern.ch/record/815187
3. CERN LHC design report v.3: the LHC injector chain. ebook: https://doi.org/10.5170/CERN-2004-003-V-3. http://cds.cern.ch/record/823808
4. ATLAS Collaboration (2008) The ATLAS experiment at the CERN large hadron collider. JINST 3:S08003. https://doi.org/10.1088/1748-0221/3/08/S08003
5. CMS Collaboration (2008) The CMS experiment at the CERN LHC. JINST 3:S08004. https://doi.org/10.1088/1748-0221/3/08/S08004
6. ALICE Collaboration (2008) The ALICE experiment at the CERN LHC. JINST 3:S08002. https://doi.org/10.1088/1748-0221/3/08/S08002

7. LHCb Collaboration (2008) The LHCb detector at the LHC. JINST 3:S08005. https://doi.org/10.1088/1748-0221/3/08/S08005
8. CMS Collaboration (2014) Description and performance of track and primary-vertex reconstruction with the CMS tracker. JINST 9:P10009. https://doi.org/10.1088/1748-0221/9/10/P10009. arXiv:1405.6569
9. CMS Collaboration (2012) CMS technical design report for the pixel detector upgrade. CMS technical design report CMS-TDR-011. https://cds.cern.ch/record/1481838
10. CMS Collaboration (2020) Track impact parameter resolution for the full pseudo rapidity coverage in the 2017 dataset with the CMS Phase-1 pixel detector. CMS detector performance note CMS-DP-2020-049. https://cds.cern.ch/record/2743740
11. CMS Collaboration (2017) The CMS trigger system. JINST 12:P01020. https://doi.org/10.1088/1748-0221/12/01/P01020. arXiv:1609.02366
12. CMS Collaboration (2020) Performance of the CMS level-1 trigger in proton-proton collisions at $\sqrt{s} = 13$ TeV. JINST 15:P10017. https://doi.org/10.1088/1748-0221/15/10/P10017. arXiv:2006.10165

Chapter 4
Data Analysis Technologies

4.1 The Particle Flow Algorithm

A crucial step towards identifying interesting events produced in pp collisions is to correctly identify and measure the numerous particles that are typically produced in such collision events. The combination of information read out from all channels of the various subdetectors of the CMS detector, to identify the type, energy, momentum, and trajectory of the final-state stable particles, that are produced as a result of pp collision events and leave an identifiable signature in at least one of the subdetectors, is referred to as physics object reconstruction. Neutrinos (and other possible BSM particles) do not leave signatures on the detectors, and are indirectly identifiable by balancing the initial and final momenta in the transverse plane. As quarks hadronise immediately after the collision, the clusters of final-state hadrons are reconstructed as jets. Heavier particles like the top quark and tau lepton as well as massive bosons are not detected directly, but their presence can usually be inferred by reconstructing their decay products. Thus, all SM particles can be detected either directly or indirectly at the CMS experiment, albeit with different efficiencies and misidentification probabilities. The final-state stable particles (electrons, muons, photons, as well as compound entities like jets, missing energy) are referred to as physics objects. This subsection discusses how each physics object is reconstructed using information from all subdetector components with the Particle Flow (PF) algorithm.

The PF algorithm [1] identifies and reconstructs individual stable particles produced in pp collision events by exploiting and combining information recorded by all the subdetectors in the CMS experiment. In addition, it estimates the missing transverse energy (MET), and is used to reconstruct the primary interaction vertex. The PF algorithm provides better energy resolutions and identification efficiencies of reconstructed particles compared to using only tracks, calorimeter clusters, or hits in the muon chambers individually.

S. Mondal, *Charming Decays of the Higgs, Z, and W Bosons*, Springer Theses,
https://doi.org/10.1007/978-3-031-84677-9_4

The PF algorithm operates in three stages, namely track and cluster finding, association, and particle identification. These steps are briefly discussed as follows:

Track and cluster finding: Tracks are reconstructed starting from pairs and triplets of hits recorded by the tracking system. This is performed using an iterative algorithm called Combinatorial Track Finder, that uses Kalman filters [2] along with pattern recognition and track fitting techniques. Energy deposits in the ECAL and HCAL are clustered separately starting with calorimeter cells that record local energy maxima as seeds for each cluster. Lastly, tracks in the muon chamber are obtained starting with hits measured by the DT or CSC chambers as seeds, followed by implementing pattern recognition algorithms on the combined readouts of the DT, CSC, and RPC chambers.

Association: A link algorithm is used to associate tracks and clusters obtained in the previous step. Tracks measured in the tracker are extrapolated to the ECAL and HCAL and are associated to a cluster if the extrapolated line lies within the cluster boundaries. When several tracks are matched to the same cluster, or several clusters are associated with the same track, only the spatially nearest combination is kept. Similarly, clusters in the ECAL are associated to the ones in the HCAL if they overlap. Finally, charged tracks in the tracker are extrapolated to the muon system, and compatible tracker tracks and muon tracks are combined into a global muon track.

Particle identification: A final particle identification is performed using the PF candidates and the associations between them as obtained in the previous steps. The procedure takes place iteratively and in a predetermined order. In each step, the identified PF candidate(s) is (are) removed from the list of PF candidates to be processed at the next step. At first, muons are reconstructed using tracker tracks matched to muon tracks, and the expected energy deposited in the calorimeters are subtracted. Next, electrons are reconstructed using tracker tracks matched to ECAL clusters. Charged hadrons are inferred from the tracker tracks matched with HCAL clusters. Finally, neutral hadrons and photons are reconstructed from HCAL and ECAL clusters, respectively, without a corresponding tracker activity. The MET of the event is calculated as the negative vector sum of the p_{T} of all identified PF objects.

The PF algorithm also helps determine the location of the primary interaction point. To do this, at first a number of high quality tracks are defined, using the number of pixel hits and the fit quality in the track fitting algorithm. Next, groups of such tracks are clustered based on the location of the z-coordinate of their closest approach to the beam pipe. This typically gives rise to several points of collisions (vertices) per bunch crossing owing to several secondary interactions. Finally, the sum of the scalar transverse momenta of all tracks matched to each vertex is evaluated, and the vertex with the highest sum is identified as the primary vertex (PV). Other vertices are considered to arise from soft scattering of other partons and PF particles associated to such vertices can be discarded as necessary.

4.2 Simulation

Simulated samples of pp collision events are useful to model expected background events as well as signal events according to SM expectations. They are also used to train several algorithms, e.g. jet flavour identification algorithms. As explained in Sect. 6.3, simulations are often prone to mismodellings, and therefore require dedicated corrections to match actual collision data. Nevertheless, they are an indispensable tool in modern particle physics analyses. This subsection briefly discusses the methods and software used in CMS for simulating physics processes.

Proton-proton collisions can be interpreted as a hard scattering process involving two partons, along with other soft interactions. The fraction of the proton momentum that a certain parton carries is probabilistic and is described by Parton Distribution Functions (PDFs), which are different for different partons and depend on the total momentum transfer, Q^2. These probabilistic scenarios are best simulated using Monte Carlo (MC) simulations. Such simulations are implemented using so-called MC generators, which account for a comprehensive set of matrix elements (MEs) and provide a fixed order perturbative QCD calculation of up to four noncollinear high-p_T hard partons. Examples of MC generators include POWHEG (V2) [3–6] and MADGRAPH5_AMC@NLO [7]. Both of these have been used to generate different physics processes used in this thesis. The PDF set NLO NNPDF3.0 [8] (NNLO NNPDF3.1 [9]) is used for the simulation of samples corresponding to the 2016 (2017 and 2018) data-taking year(s).

In addition to the ME generation, soft and collinear radiations are described using a parton showering (PS) approach. The PS tool used in the simulation of all samples used in the presented analyses is PYTHIA8 [10, 11]. Whenever there is an overlap of the phase space between the calculation of the ME and the simulation of the PS, so-called matching algorithms are used to resolve duplicates. Examples of matching algorithms are MLM [12] and FXFX [13]. The former has been used in conjuction with processes simulated at the leading order (LO), while the latter is used with those generated at the next-to-leading order (NLO) in QCD.

Besides the hard scattering of the partons, remnants of the colliding proton may interact with particles from the hard scattering, multiple partons can interact in the same pp collision, or multiple protons in the same bunch crossing can collide, giving rise to additional pileup events. All these effects are termed as the underlying event. The effect of additional pp interactions within the same or nearby bunch crossings on top of the hard scattering processes is modelled by additional minimum bias collisions generated with PYTHIA. The underlying event tune used in the presented analyses for processes involving top quarks is CP5 [14]. For other processes, the CUETP8M1 (CP5) tune is used for the samples corresponding to the 2016 (2017 and 2018) data-taking year(s).

Finally, the interactions of all the final-state stable particles generated in the previous steps with the CMS detector are simulated using GEANT4 [15]. The resulting readouts of all the channels of all the subdetectors are simulated. These responses effectively resemble the actual detector response, had the same event been recorded in a real pp collision. Next, the same reconstruction algorithms, that are used to reconstruct events from real pp collisions, are applied to the simulated data and the outputs are stored as simulated datasets.

The simulated datasets can technically be as large as needed. Given the entire simulation chain is computationally quite intensive, the number of simulated events practically range between tens of thousands to billions per physics process. The event yields of each of these processes after applying analysis-specific criteria are rescaled to reflect the theoretical or measured cross section, σ, of the given process, and the integrated luminosity, $\mathcal{L}$, of the corresponding data. If an event, i, of a given process has a generated weight of $w_{\mathrm{gen,i}}$, the event would be assigned an effective weight of

$$w_i = \frac{\sigma \cdot \mathcal{L} \cdot w_{\mathrm{gen},i}}{\sum_j^N w_{\mathrm{gen},j}}, \tag{4.1}$$

where N is the total number of events generated for the sample before any selection criteria are applied. This, however, does not include any additional corrections that are usually applied to account for mismodellings in the simulation; such additional weights are described in Sect. 9.7.

4.3 CMS Data Tiers

The CMS experiment stores data and simulated samples in the so-called Event Data Model (EDM) format, centering around the concept of an Event. An Event is a C++ object containing all raw (RAW) and reconstructed data related to a particular collision. It starts as a collection of RAW data from the detector or a simulated event, stored as a single entity in memory called `edm::Event`. As the event data is processed, products of every processing step are stored in the Event as reconstructed (RECO) data objects. The Event thus holds all data that was taken during a triggered physics event as well as quantities derived from the raw data during the reconstruction steps. The Event also contains metadata describing the configuration of the software used for the reconstruction of each physics object and the conditions and calibration data used for such reconstruction.

However, for physics analysis purposes, often a small subset of the information present in the RECO data format is sufficient. For storage and processing efficiencies, the Analysis Object Data (AOD) tier was developed as a subset of RECO for all analysis purposes. Further, an even smaller subset, the MiniAOD data tier, was introduced in 2014 to facilitate faster analysis and more efficient storage of data.

Finally, a compact, more user-friendly data tier called the NanoAOD was introduced in 2017. NanoAOD is not an EDM format, but has a simpler tree-like structure that can be read using ROOT (cf. Sect. 4.5). It contains per-event information that is most commonly used by a majority of analyses. The size per event of NanoAOD is of the order of $\sim$ 1 KB, making storage and readout very convenient. NanoAOD files can be generated from their corresponding MiniAOD files, and are usually generated centrally by the CMS collaboration as often as necessary. Individuals in the collaboration can also privately produce custom NanoAOD samples with additional information from MiniAOD, if necessary. The format is still being improved based on additional feature requests to accommodate more commonly-used event information. The guiding principle behind this data tier is to reduce file size as far as practicable, and therefore derived quantities are often not included. Such quantities can be recalculated as necessary using certain frameworks, such as the NANOAOD POSTPROCESSOR [16], also maintained centrally by the collaboration.

The analyses discussed in this thesis make use of the NanoAOD data format. The final set of samples used were custom NanoAOD samples with certain additions compared to the centrally produced ones. Further, the NANOAOD POSTPROCESSOR tool was used to compute certain derived quantities from the ones stored in the NanoAOD format.

4.4 Boosted Decision Trees, Neural Networks, and Deep Learning

Artificial Neural Networks, usually referred to as Neural Networks (NNs), are computational systems used in Multivariate Analysis (MVA) problems that use a network of functions to understand and map a data input, usually with a large number of dimensions, to a data output, usually with a smaller number of dimensions. The concept is inspired by the way neurons of the human brain function and work together to process complex information from sense organs. NNs are an example of Machine Learning (ML), which in turn is a subset of the broader field of Artificial Intelligence (AI). Deep Learning (DL) based on Deep Neural Networks (DNNs) [17] extends the concept of NNs by utilising multiple layers of processing that extract progressively higher level features from the input data. NNs and DNNs constitute a broad field of research today, with numerous applications across disciplines. This subsection provides a brief outline of the key concepts that have been used in the presented analyses.

Implementation of NNs usually involves "training" a network to recognise patterns in an input dataset to develop a function that maps each input x to its corresponding output y. The process involves developing a model that consists of a vast number of free parameters and maps the inputs to the outputs. Supervised learning refers to such a training performed with known input-output pairs. The process can be viewed as a usual minimisation task, where a function that quantifies the difference between the predicted output and the true output is minimised progressively in several "epochs". Such a function is usually referred to as the "loss function", and its functional form and behaviour is often specific to the task at hand. The training is considered to have converged if the loss value cannot be further minimised. Once the training converges, the model with frozen parameters can be used to predict the outputs of a different set of inputs whose outputs are unknown.

Often an additional test step is involved after the training step where one checks the validity of the model on an independent set of known input-output pairs . A similar value of the loss function for both the training and test datasets ensures that the model can be generalised to populations independent of the input training dataset. A higher value of the loss function in the test set is a signature of "overtraining", which usually means the model has picked up unique features of the input dataset that are not general to the entire population. Such overtraining is avoided using several techniques, such as using a larger training dataset, or tuning the so-called hyperparameters of the training process. These hyperparameters are training configurations that cannot be deduced from the dataset, but are provided externally before the training is performed. Therefore, tuning these hyperparameters to mitigate overtraining, if any, is usually performed on a trial-and-error basis, or with random or sequential search algorithms.

Another approach in MVA, known as decision trees (DTs), involves recognising patterns in the input dataset by iteratively partitioning data by asking questions ("if-then-else" comparisons) about the input data features, much akin to flowcharts. Learning on the basis of such conditional control statements to make decisions is referred to as "weak learning". Boosted Decision Trees (BDTs) [18] involves combining several such DTs in series to achieve a strong learner from many sequentially connected weak learners. In each step, the algorithm filters out observations that a learner gets correct, thereby stressing on adding new weak learners that are capable of handling progressively more complex observations in every step. The usage of BDTs is very similar to that of NNs and DNNs, involving a training step, followed by testing, and finally application on data for which predictions are sought.

The most common tasks in which NNs and BDTs are implemented are classification and regression. Classification tasks involve discrete outputs and are useful when inputs need to be labelled as different qualitative classes. Classification tasks are relevant in this thesis in the context of jet flavour tagging and signal-event identification, for example. On the other hand, regression tasks involve outputs that are continuous, real numbers. Examples include finding the best estimate of the energy of the quark that initiated a jet, based on the kinematics of the jet's constituents.

4.5 Software and Packages

The several software, packages, frameworks, and libraries used in performing the analyses discussed in this thesis are listed below with a short description for each.

CMSSW: The overall collection of software used at the CMS experiment is referred to as CMSSW [19]. It contains the services needed by the simulation, calibration and alignment, and reconstruction modules that process event data so that physicists can perform analysis. The framework facilitates the development and deployment of reconstruction and analysis software.

ROOT: ROOT [20] is a software framework for data analysis developed by CERN. It works as an interpreter based on a C++ backend. It facilitates storage of complex data in formats suitable for analyses in experimental particle physics. It also offers object-oriented classes corresponding to the several reconstructed physics objects with pre-programmed methods that aid standard mathematical operations on physical observables. In addition, ROOT offers a number of data visualisation and plotting tools. ROOT can be interfaced with the programming language, Python, using the PyROOT library developed for Python. This provides additional flexibility to switch between C++ and Python for analysis codes.

RDataFrame: RDataFrame [21] is a relatively new library in the ROOT environment that offers a modern, high-level interface for analysis of data stored in ROOT formats. It interprets data in the form of so-called dataframes, which are tabular structures of data, and provides an intuitive and convenient way to perform operations. It is based on a "lazy" action approach, which means the actual loop execution on data is deferred until at least one output is explicitly requested. A user can, therefore, define multiple chains of operations and transformations on the same dataset without looping through the data every time; instead the event loop is triggered only once and all actions are performed in the same loop, returning all results at the same time at the end of the loop. The implementation also automatically puts in place several low-level optimisations such as multi-thread parallelisation and caching.

TMVA: TMVA [22] is a ROOT library that provides the interfaces and implementations of several ML techniques. This includes NNs, DNNs, and BDTs. It also provides plotting tools to inspect the performance of the trained models. The TMVA library has been used to train BDT-based classifiers to separate signal from background in the VHcc and VZcc analyses (cf. Sect. 10.1). Among the various BDT algorithms available in TMVA, the "gradient boost" algorithm is used.

NanoAOD Postprocessor: The NanoAOD Postprocessor [16] is a data processing tool maintained centrally by the CMS collaboration. It provides tools to add derived quantities to the list of quantities already existing in a NanoAOD file. For example, while jet p_Ts are stored in NanoAOD, the systematic variations of jet energy calibrations are not stored. The Postprocessor tool uses the raw p_T of a jet along with the calibration information provided externally to recalculate and append all jet energy systematic variations. Other tasks that can be accomplished with the Postprocessor tool are calculating pileup weights of simulated events based on externally-provided pileup distributions observed in data, evaluating flavour tagging weights based on jet configurations in an event, etc.

AnalysisTools: AnalysisTools is a C++-based event processor developed originally for the VH(H $\rightarrow$ b$\bar{\text{b}}$) (VHbb) analysis [23]. A version of the same, adapted for the VHcc analysis, performs the bulk of the event processing tasks required in the analysis. This includes calculating complex observables, applying selections, categorising events into channels and regions, applying corrections to simulations, evaluating BDT predictions, evaluating systematic uncertainties, etc.

ShapeMaker: ShapeMaker is a tool first developed for the VHcc analysis [24] for plotting histograms with all systematic variations that are required for the final statistical analysis. It implements the novel RDATAFRAME library [21] to process the outputs of ANALYSISTOOLS and generate histograms for every process, every channel and region, with every systematic variation, in a multithreaded implementation. It dramatically reduces disk access and processing times compared to traditional methods, reducing the runtime of the shape generation workflow from the order of ~a week to the order of ~a few hours, with a fraction of the CPU usage. The framework also has features to reevaluate certain observables, apply additional selections and corrections, and reevaluate BDTs on the run, as necessary.

SCIPY: SCIPY [25] is a Python library used for scientific and technical computing. It contains modules for optimization, linear algebra, integration, interpolation, special functions, FFT, signal and image processing, ODE solvers and other tasks common in science and engineering. In this thesis, a number of its submodules have been implemented in the context of interpolating charm tagger calibration factors and deriving interpolation uncertainties.

Other packages: GNU `parallel` [26] is a shell tool for executing jobs in parallel on a computer and has been used to run some codes in the course of the thesis. A part of the SCIKIT- HEP project [27, 28], `mplhep` [29] has been used in conjunction with MATPLOTLIB [30] on Python to produce certain plots in this thesis.

4.6 Remote Computing

The analyses presented in this thesis involve analysing large volumes of data stored in computing systems across the world. It is far beyond the scope of the storage, memory, and processing capacities of the computing systems available locally to accommodate and process such volumes of data. This problem is overcome by running computational jobs on remote computing clusters, that have the required data available locally or have achitectures enabling quick access to data stored elsewhere. For this purpose, the CMS experiment has several so-caller Tier-2 (T2) storages available across the globe. Each of these T2 centres usually have a local computing cluster that has direct and fast read access of the T2 storage. In this thesis, the storage and computing systems of the DESY and RWTH T2 centres were used. In addition, the EOS file system and the LXPLUS computing system at CERN were also used. The Secure Shell (`ssh`) protocol was used to remotely log in to the remote clusters to submit computational tasks.

For the purpose of this analysis, the number of core hours required to process the reconstructed data and simulation made available centrally by the CMS experiment, not including the time required to produce such data, is of the order of $\sim$ 1 million. It is clearly not feasible to run one single process to analyse all detector and simulation data sequentially. Therefore, several processors, contained in so-called "batch systems", are deployed to access and process the data in parallel. Each of the processors handling a "job" processes anywhere between 1000–1,000,000 events depending on the complexity of the job. As every job can be processed independently of other jobs, no information needs to be exchanged between two peer processors during the course of the run. This method of accomplishing a large task using several computing nodes is known as High Throughput Computing (HTC).

HTC has been implemented in this thesis using tools like CRAB3 [31] and HTCondor [32]. CRAB3 is a dedicated utility developed and maintained centrally by the CMS collaboration to submit CMSSW jobs to distributed computing resources across the world. The submitted jobs are sent to specified sites, preferably at sites where the input data is locally available, where they run remotely. Protocols like XRootD [33] also allow direct access to files stored in remote sites. Once the job is complete, the generated outputs are transferred back to a user-specified site. The NanoAOD PostProcessor step for the presented analyses has been performed using CRAB3.

HTCondor, on the other hand, distributes jobs on a local cluster and is capable of running any executable file compatible with the installed operating system. The HTCondor installations available on the National Analysis Facility (NAF) computing system at DESY and the LXPLUS computing system at CERN were used in conjunction with the DESY T2 storage and EOS file system, respectively, to implement HTC during the course of this thesis. The AnalysisTools and ShapeMaker steps of the analysis, as well as the charm-tagger calibration workflow, were executed using HTCondor.

4.7 Statistical Tools

This subsection briefly outlines some concepts of statistical analysis that have been implemented in this thesis.

4.7.1 ROC Curves

A Receiver Operating Characteristic (ROC) curve is a graphical plot that illustrates the classification accuracy in a binary classification system at varied thresholds of its discrimination. It is useful in judging the performance of a signal-background discriminator, for instance a jet flavour identification algorithm.

A True Positive (TP) refers to a prediction of a binary classifier that correctly identifies the presence of a signal. On the contrary, a False Positive (FP) refers to a prediction in which the classifier incorrectly indicates that a signal is present. The fraction of times when each of the TP and FP occur when the classifier is presented with signal and background events, respectively, are called the TP rate (TPR) and FP rate (FPR), respectively. Both the TPR and FPR rates vary as the threshold of the discriminator output is changed. This allows us to plot the TPR as a function of the FPR, or vice versa, where each rate is calculated at a fixed discriminator threshold. Such a plot is known as the ROC curve of the discriminator. For example, if the background is denoted as 0 and signal as 1, and a "presence" is inferred every time the classifier predicts an output value (discriminator) of p ($p \in [0, 1]$) or above, one gets a pair of TPR and FPR values for a fixed value of p. Sliding the value of p between 0 and 1, one can obtain a series of [TPR,FPR] pairs, which can be plotted to obtain a ROC curve.

A random classifier would always yield outputs with TPR=FPR, thereby producing a ROC curve that is a straight line connecting (0,0) and (1,1). This corresponds to an area under the curve (AUC) of 0.5 and represents a trivial case. On the other hand, a perfect classifier would be just a point on the ROC curve at TPR=1 and FPR=0. A realistic classifier would be a curved line connecting (0,0) and (1,1) with an AUC between 0.5 and 1 (or 0.5 and 0, depending on which axis is plotted on the x-axis). The further away from 0.5 the AUC is, the more performant the classifier is.

In this thesis, ROC curves have been presented in the context of jet flavour identification performance (Sects. 6.2 and 7.7) and BDT training for signal vs. background discrimination of the VHcc and VZcc analyses (Sect. 10.1).

4.7.2 Binned Maximum Likelihood Fit

A common method to detect presence (or absence) of signal in experimental particle physics analyses is the method of maximum likelihood estimation. The method involves performing a statistical hypothesis test with the background- (B) only model as the null hypothesis and the signal-plus-background (S+B) model as the alternative hypothesis. Often, the signal is scaled by a free parameter μ, described as a signal strength modifier, and the best-fit value of μ is estimated from the available data. This makes μS+B the alternative hypothesis. The Maximum Likelihood Estimation (MLE) method provides a prescription for testing whether the observed collision data favours one model over another. According to the Neyman-Pearson lemma [34], the most powerful test statistic to evaluate two competing hypotheses is based on the ratio of the likelihoods of the observation under the two hypotheses.

In the analyses discussed in this thesis, a binned maximum likelihood fit is performed, which means the data is modelled in several histogram bins as a function of a certain discriminating variable. Given that for a single counting experiment, the probability to find x events, when the predicted number of events is λ, follows a Poisson distribution

$$\mathrm{Poi}(x|\lambda) = \frac{\lambda^x}{x!}e^{-\lambda}, \tag{4.2}$$

the same for N bins with λ_i expected in each, would be given by

$$\mathcal{P}(x|\lambda) = \prod_i^N \mathrm{Poi}(x_i|\lambda_i). \tag{4.3}$$

With λ_i simulated from signal and background samples, the probability of observing x_i events under the alternative hypothesis ($\lambda_i = \mu s_i + b_i$) is, therefore,

$$\mathcal{P}(x|\mu S + B) = \prod_i^N \text{Poi}(x_i|\mu s_i + b_i), \tag{4.4}$$

where s_i and b_i are the predicted signal and background yields, respectively, in bin i. However, s_i and b_i have associated uncertainties, called nuisance parameters (NPs), modelled as statistical and systematic uncertainties and summarised in Sect. 9.8. Since NPs, $\boldsymbol{\theta}$, allow signal and background yields to fluctuate from nominal values in each bin, s_i and b_i can be rewritten as $s_i(\boldsymbol{\theta})$ and $b_i(\boldsymbol{\theta})$, respectively. For each systematic, j, an additional penalty term, $\pi_j(\boldsymbol{\theta})$ can be associated to the likelihood function. Therefore, the likelihood function takes the form

$$L(x|\mu, \boldsymbol{\theta}) = \prod_i^N \text{Poi}(x_i|\mu s_i(\boldsymbol{\theta}) + b_i(\boldsymbol{\theta})) \times \prod_j^{N_s} \pi_j(\boldsymbol{\theta}), \tag{4.5}$$

with N_s as the total number of NPs, including the bin-by-bin statistical uncertainties in the prediction. Depending on the type of uncertainty, $\pi_j(\boldsymbol{\theta})$ may apply to all bins equally, or to just one bin (e.g. for bin-by-bin statistical uncertainties), or to different bins with different magnitudes. Similarly, $\pi_j(\boldsymbol{\theta})$ may apply to just one, several, or all physics processes involved in the signal and background compositions. Depending on the expected behaviour, the NPs can be classified into three types:

Log-normal (rate changing) uncertainties: NPs that change only the rates of processes but not the shape of the templates (that is relative bin-by-bin contributions) are assigned as log-normal uncertainties. These uncertainties are usually evaluated as Gaussian distributions. However, for large uncertainties, the -1σ shift (that is, the lower end of the 68% confidence interval) of the Gaussian may potentially assume negative values and therefore must be truncated. Hence, a log-normal distribution is assigned instead.

Shape uncertainties: Certain nuisances are correlated across the bins of the fit variable and therefore affect the shape of the templates being fitted. These shape variations are taken into account by producing separate templates for the $\pm 1\sigma$ (that is, upper and lower limits of the 68% confidence interval) in addition to the nominal template.

Bin-by-bin (statistical) uncertainties: Since the number of events in each bin of the histograms is finite, a statistical uncertainty is associated with every bin. This NP is usually approximated using a Gaussian distribution per bin, instead of a Poisson uncertainty for each process in each bin. This is a reasonable approximation for a suitably large number of simulated events per bin [35]. To this end, a dedicated rebinning of the histograms is performed to ensure sufficient number of events in the fitted distributions, especially at the tails.

Once the likelihood function is defined, one can construct a test statistic based on the likelihood ratio. For a given signal strength μ, the test statistic $\tilde{q}(\mu)$ takes the form

$$\tilde{q}(\mu) = -2\ln\left[\frac{L\left(x|\mu, \hat{\theta}_{\mu}\right)}{L\left(x|\hat{\mu}, \hat{\theta}\right)}\right], \tag{4.6}$$

where the numerator $L\left(x|\mu, \hat{\theta}_{\mu}\right)$ is the likelihood in Eq. 4.5 evaluated at the given μ at its maximum (which translates to $\hat{\theta}_{\mu}$ being the best-fit $\hat{\theta}$ for μ), and the denominator $L\left(x|\hat{\mu}, \hat{\theta}\right)$ is the likelihood evaluated at its global maximum (at the joint best-fit $\hat{\mu}, \hat{\theta}$). For a given measurement of the observed parameter, x, one obtains a single value of $\tilde{q}(\mu)$. However, performing the hypothesis test requires knowing the probability distribution $f\left(\tilde{q}(\mu)|\mu\right)$ of $\tilde{q}(\mu)$ for a given outcome, μ. According to Wilk's theorem [36], for sufficiently large number of bins, $\tilde{q}(\mu)$ asymptotically converges towards a χ^2 distribution. Therefore, asymptotic formulae can be used to derive $f\left(\tilde{q}(\mu)|\mu\right)$ analytically [37]. Depending on the signal strength, one can either measure the "significance" of the signal, or can interpret the result as an "upper limit".

Significance In case of a non-zero signal strength, $\hat{\mu}$, one can quantify the significance of the signal by comparing the value of the resulting test statistic $\tilde{q}(\hat{\mu})$ with the probability distribution of the test statistic under the background-only ($\mu = 0$) hypothesis ($f\left(\tilde{q}(0)|0\right)$). This can be done, for example, using the p-value of the observation given the background-only hypothesis:

$$\text{p-val}(\text{obs}|\mu = 0). \tag{4.7}$$

This p-value can be interpreted as the probability of the observation being just a statistical fluctuation consistent with the background-only hypothesis. The p-value can also be translated to the number (N) of standard deviations (σ) an observation is away from the mean of a normal distribution, commonly termed as "$N\sigma$ (N-sigma) significance". A significance of 3σ is considered as an evidence for a process, while a significance of 5σ is considered an observation in particle physics.

Upper limit In case the best-fit signal strength is less than 3σ, the result is usually interpreted as a statistical fluctuation consistent with the background-only hypothesis. One can, however, calculate an upper limit on the signal strength, above which the signal is excluded at, for example, 95% confidence. The CL_S method [37–39] that is generally used in such cases involves scanning over various values of the signal strength and calculating the p-value of obtaining such a signal under the signal-plus-background hypothesis. However, such a metric is not robust against downward fluctuations of the measurement, and therefore the CL_S metric is defined as the ratio of two p-values, as

$$\text{CL}_\text{S} = \frac{\text{p-val}(\text{obs}|\mu S + B)}{1 - \text{p-val}(\text{obs}|B)}, \tag{4.8}$$

where the p-val in the denominator represents the p-value of the observation under the background-only hypothesis. The values of μ, for which $\text{CL}_\text{S} < 0.05$, are considered excluded by the data at a 95% confidence level. The lower threshold of such excluded values is denoted as $\mu_{0.05}$, so that $\mu > \mu_{0.05}$ denote excluded signal-strength values.

In particle physics analyses, the standard procedure is to first "blind" the signal regions, that is, one does not look at the agreement between data and simulation in parts of the search phase space where a large fraction of the signal is expected. The motivation behind doing this is to make sure one does not introduce a subconscious bias in the simulation-based data modelling that may potentially lead to making observations based on what one expects, instead of what the data really quantifies. Therefore, the data in the signal regions is "unblinded" only when the background modelling based on the control regions looks reasonable. One can also calculate an "expected limit" based on the blinded simulation. This is done by creating numerous "toy" datasets from the simulated background-only models by letting the bin values shift away from the nominal yield as per the descriptions of all the NPs. Following this, the CL_S and $\mu_{0.05}$ values for each of these toy datasets are calculated just as one would do with the real observation. This gives a probability distribution of $\mu_{0.05}$ values. The median of these distributions (that is, with 50% cumulative probability) is termed as the median expected upper limit of the signal, hereafter referred to as simply the expected upper limit. One can also obtain $\pm 1\sigma$ and $\pm 2\sigma$ bands of this expectation by looking at the distribution of $\mu_{0.05}$ values, thereby quantifying the fluctuations of the expected upper limit within the standard 68% and 95% confidence intervals, respectively. It can be noted that the confidence intervals on the distribution of the expected $\mu_{0.05}$ is not to be confused with the threshold on CL_S and the exclusion of signal-strengths at the 95% confidence level.

Impacts The effect of each NP on the best fit signal strength $\hat{\mu}$ can be evaluated by once again constructing numerous toy datasets with varied systematic values (θ_j) sampled from their respective distributions. For each of these toy models, one can find the best-fit $\hat{\mu}$ and best-fit $\hat{\theta}_j$ by performing the usual fits. The changes in the value of the best-fit $\hat{\mu}$ due to the variation of a systematic θ_j gives an estimate of the "impact" of the said systematic on the best-fit $\hat{\mu}$ value. The changes in the value of the best-fit $\hat{\theta}_j$ from that of the pre-fit θ_j, in units of the probability distribution widths of θ_j, give the so-called "pull" of the systematic.

In this thesis, all statistical approaches towards signal extraction have been implemented using the CombineHarvester [40] and Higgs Combine [41] tools, which are used across several physics analyses in CMS. Statistical tools in ROOT (cf. Sect. 4.5), such as `RooStats` and `RooFit`, form the basis of these frameworks.

References

1. CMS Collaboration (2017) Particle-flow reconstruction and global event description with the CMS detector. JINST 12:P10003. https://doi.org/10.1088/1748-0221/12/10/P10003. arXiv:1706.04965
2. Kalman RE (1960) A new approach to linear filtering and prediction problems. J Basic Eng 82:35–45. https://doi.org/10.1115/1.3662552
3. Nason P (2004) A new method for combining NLO QCD with shower Monte Carlo algorithms. JHEP 11:040. https://doi.org/10.1088/1126-6708/2004/11/040. arXiv:hep-ph/0409146
4. Frixione S, Nason P, Oleari C (2007) Matching NLO QCD computations with parton shower simulations: the POWHEG method. JHEP 11:070. https://doi.org/10.1088/1126-6708/2007/11/070. arXiv:0709.2092
5. Alioli S, Nason P, Oleari C, Re E (2010) A general framework for implementing NLO calculations in shower Monte Carlo programs: the POWHEG BOX. JHEP 06:043. https://doi.org/10.1007/JHEP06(2010)043. arXiv:1002.2581
6. Campbell JM, Ellis RK, Nason P, Re E (2015) Top-pair production and decay at NLO matched with parton showers. JHEP 04:114. https://doi.org/10.1007/JHEP04(2015)114. arXiv:1412.1828
7. Alwall J et al (2014) The automated computation of tree-level and next-to-leading order differential cross sections, and their matching to parton shower simulations. JHEP 07:079. https://doi.org/10.1007/JHEP07(2014)079. arXiv:1405.0301
8. NNPDF Collaboration (2015) Parton distributions for the LHC Run II. JHEP 04:040. https://doi.org/10.1007/JHEP04(2015)040. arXiv:1410.8849
9. NNPDF Collaboration (2017) Parton distributions from high-precision collider data. Eur Phys J C 77:663. https://doi.org/10.1140/epjc/s10052-017-5199-5. arXiv:1706.00428
10. Sjöstrand T et al (2015) An introduction to PYTHIA 8.2. Comput Phys Commun. 191:159. https://doi.org/10.1016/j.cpc.2015.01.024. arXiv:1410.3012
11. Sjostrand T, Mrenna S, Skands PZ (2008) A brief introduction to PYTHIA 8.1. Comput Phys Commun 178:852–867. https://doi.org/10.1016/j.cpc.2008.01.036. arXiv:0710.3820
12. Alwall J et al (2008) Comparative study of various algorithms for the merging of parton showers and matrix elements in hadronic collisions. Eur Phys J C 53:473. https://doi.org/10.1140/epjc/s10052-007-0490-5. arXiv:0706.2569
13. Frederix R, Frixione S (2012) Merging meets matching in MC@NLO. JHEP 12:061. https://doi.org/10.1007/JHEP12(2012)061. arXiv:1209.6215
14. Skands P, Carrazza S, Rojo J (2014) Tuning PYTHIA 8.1: the Monash, (2013) tune. Eur Phys J C 74:3024. https://doi.org/10.1140/epjc/s10052-014-3024-y. arXiv:1404.5630
15. GEANT4 Collaboration (2003) GEANT4-a simulation toolkit. Nucl Instrum Meth A 506:250. https://doi.org/10.1016/S0168-9002(03)01368-8
16. CMS Collaboration NanoAOD-tools. https://twiki.cern.ch/twiki/bin/view/Sandbox/NanoAODToolsTMP
17. Goodfellow I, Bengio Y, Courville A (2016) Deep learning. Adaptive computation and machine learning. MIT Press. ISBN: 9780262035613
18. Drucker H, Cortes C (1995) Boosting decision trees. In: Touretzky D, Mozer M, Hasselmo M (eds) Advances in neural information processing systems, vol 8. MIT Press
19. CMS Collaboration CMSSW. https://github.com/cms-sw/cmssw
20. Brun R, Rademakers F (1997) ROOT: an object oriented data analysis framework. Nucl Instrum Meth A 389:81–86. https://doi.org/10.1016/S0168-9002(97)00048-X
21. Piparo D et al (2019) RDataFrame: easy parallel ROOT analysis at 100 threads. EPJ Web Conf 214:06029. https://doi.org/10.1051/epjconf/201921406029
22. Therhaag J (2010) TMVA toolkit for multivariate data analysis in ROOT. PoS ICHEP2010 510. https://doi.org/10.22323/1.120.0510
23. CMS Collaboration (2018) Observation of Higgs boson decay to bottom quarks. Phys Rev Lett 121:121801. https://doi.org/10.1103/PhysRevLett.121.121801. arXiv:1808.08242

24. CMS Collaboration (2020) A search for the standard model Higgs boson decaying to charm quarks. JHEP 03:131. https://doi.org/10.1007/JHEP03(2020)131. arXiv:1912.01662
25. Virtanen P et al (2020) SciPy 1.0: fundamental algorithms for scientific computing in Python. Nat Methods 17:261. https://doi.org/10.1038/s41592-019-0686-2
26. Tange O GNU parallel. https://doi.org/10.5281/zenodo.4454976
27. Heinrich L, Feickert M, Stark G pyhf: v0.7.0. https://github.com/scikit-hep/pyhf/releases/tag/v0.7.0. https://doi.org/10.5281/zenodo.1169739
28. Heinrich L, Feickert M, Stark G, Cranmer K (2021) pyhf: pure-python implementation of histfactory statistical models. J Open Source Softw 6:2823. https://doi.org/10.21105/joss.02823
29. Novak A et al mplhep 0.3.26. https://doi.org/10.5281/zenodo.6807166. https://pypi.org/project/mplhep/
30. Hunter JD (2007) Matplotlib: a 2D graphics environment. Comput Sci & Eng 9:90–95. https://doi.org/10.1109/MCSE.2007.55
31. Mascheroni M et al (2015) CMS distributed data analysis with CRAB3. J Phys Conf Ser 664:062038. https://doi.org/10.1088/1742-6596/664/6/062038
32. HTCondor. https://htcondor.org/
33. XRootD. http://xrootd.slac.stanford.edu/
34. Neyman J, Pearson ES (1933) On the problem of the most efficient tests of statistical hypotheses. Phil Trans R Soc Lond A 231:289–337. https://doi.org/10.1098/rsta.1933.0009
35. Barlow RJ, Beeston C (1993) Fitting using finite Monte Carlo samples. Comput Phys Commun 77:219–228. https://doi.org/10.1016/0010-4655(93)90005-W
36. Wilks SS (1938) The large-sample distribution of the likelihood ratio for testing composite hypotheses. Ann Math Stat 9:60–62. https://doi.org/10.1214/aoms/1177732360
37. Cowan G, Cranmer K, Gross E, Vitells O (2011) Asymptotic formulae for likelihood-based tests of new physics. Eur Phys J C 71:1554. https://doi.org/10.1140/epjc/s10052-011-1554-0. arXiv:1007.1727. [Erratum: Eur Phys J C 73:2501 (2013)]
38. Read AL (2002) Presentation of search results: The CL_S technique. J Phys G 28:2693. https://doi.org/10.1088/0954-3899/28/10/313
39. Junk T (1999) Confidence level computation for combining searches with small statistics. Nucl Instrum Meth A 434:435. https://doi.org/10.1016/S0168-9002(99)00498-2. arXiv:hep-ex/9902006
40. CMS Collaboration Combine harvester. https://cms-analysis.github.io/CombineHarvester/
41. CMS Collaboration Higgs combine. https://cms-analysis.github.io/HiggsAnalysis-CombinedLimit/

Chapter 5
Physics Object Definitions

The PF algorithm discussed in Sect. 4.1 reconstructs physics objects with high efficiencies. However, physics analyses often require tighter quality criteria for certain physics objects in order to further suppress backgrounds that contain other physics objects misidentified as objects of interest. This, of course, comes at the cost of slightly lower efficiencies. Furthermore, certain additional corrections are often applied to improve the resolution of energy measurements of certain particles. This section discusses the additional corrections and quality criteria that have been used to define physics objects used in the presented analyses.

5.1 Muons

As discussed in Sect. 4.1, the PF algorithm reconstructs muons using information from the tracker as well as the muon chambers. This yields a very high efficiency of muon identification within the η range of the muon chambers ($|\eta| < 2.4$) [1, 2]. A calibration of the muon momenta is performed using the "Rochester correction method" to compensate for mismeasured momenta due to detector misalignment and other experimental imperfections [2]. These corrections are derived by comparing distributions of simulated $Z/\gamma^* \to \mu^+\mu^-$ samples with a selection of data enriched in such processes, and extracting correction factors in granular bins of p_T, η, and ϕ.

A number of additional quality and kinematic criteria are imposed on the PF muons used for the analyses discussed in this thesis, depending on the purpose of muon selection. The CMS Muon Physics Object Group (POG) defines some of these criteria for common usage across all CMS analyses. For instance, "loose muon" refers to all PF muons, while the "tight muon" refers to a subset of PF muons that have a good fit quality in the global-muon track fit as well as a high number of tracker hits and a low transverse IP $|d_{xy}| < 2$ mm with respect to the PV [2]. Another criterion, referred to as relative isolation ($r_{\text{iso}}^{\Delta R}$), is defined as the scalar p_T sum of the PF candidates

S. Mondal, *Charming Decays of the Higgs, Z, and W Bosons*, Springer Theses,
https://doi.org/10.1007/978-3-031-84677-9_5

within a cone of radius, $\Delta R = \sqrt{(\Delta\eta)^2 + (\Delta\phi)^2}$, around the muon divided by the muon p_T, where ϕ is the azimuthal angle in radians. This criterion distinguishes prompt muons, originating from the hard scattering process, from secondary muons, originating, for example, from semileptonic decays of heavy hadrons. The latter type of muons usually have low p_T (soft) and have hadronic activity surrounding them, resulting in a high value of $r_{\text{iso}}^{\Delta R}$.

The muon selections used in the analyses described in this thesis can be categorised into six categories as follows:

1. Isolated, prompt muons for charm tagger calibration: Tight muons with $r_{\text{iso}}^{0.4} < 0.05$, $|d_{xy}| < 0.002$ cm, and longitudinal IP $|d_z| < 0.01$ cm.
2. Isolated, prompt muons for charm tagger calibration in dileptonic channels: Tight muons with $r_{\text{iso}}^{0.4} < 0.15$.
3. Non-isolated, soft muons for charm tagger calibration: Tight muons with $r_{\text{iso}}^{0.4} > 0.2$ (or $r_{\text{iso}}^{0.4} > 0.5$) and p_T between 5 and 25 GeV.
4. Isolated, prompt muons in 1L channel: Tight muons with $p_T > 25$ GeV, $r_{\text{iso}}^{0.4} < 0.06$, $|d_{xy}| < 0.05$ cm, and $|d_z| < 0.2$ cm.
5. Isolated, prompt muons in 2L channel: Loose muons with $p_T > 20$ GeV and $r_{\text{iso}}^{0.4} < 0.25$.
6. Veto muons in 0L and 1L channels: Loose muons with $p_T > 15$ GeV and $r_{\text{iso}}^{0.4} < 0.1$.

The usage of various types of muon selections is elaborated in Sects. 7.2 and 9.3. The p_T thresholds of the reconstructed muons are generally motivated by the p_T thresholds of the HLT paths corresponding to the selections they are used in.

5.2 Electrons

The selection of electron candidates starts with the PF identification as explained in Sect. 4.1. Thereafter, the CMS EGAMMA POG applies a multivariate electron identification method to select a subset of PF electron candidates which correspond to high-quality, prompt electrons [3, 4]. This is done by applying MVA methods on electrons selected from simulated Drell–Yan (DY)+Jets samples, with prompt electrons matched to Z decays as signal, and unmatched and non-prompt electrons as background. Using a standard BDT-based approach, the POG obtains a single discriminator variable, that is computed based on multiple parameters of the electron object and provides the best separation between the signal and backgrounds. An analyzer can, in principle, require a selection criterion on any value of this BDT and choose electrons that pass this criterion—the so-called working point (WP). However, the EGAMMA POG recommends two WPs, corresponding to 80% (tighter) and 90% (looser) electron efficiencies, respectively, as WPs to be used in analyses across CMS [4]. These are referred to as the MVA IDs. Corrections to simulations to match the behaviour in data are made available centrally at CMS for these two WPs.

Furthermore, a relative isolation $r_{\text{iso}}^{\Delta R}$ can be defined for electrons, similar to muons, to distinguish prompt electrons from secondary ones. Electron identification is performed within the η range of the tracker. However, electrons reconstructed within $1.442 < |\eta| < 1.556$ are discarded as these include electrons traversing through the gap between the barrel and endcap regions of the ECAL.

Depending on the usage, five types of electron selections have been used in this thesis:

1. Isolated, prompt electrons for charm tagger calibration: MVA ID with 80% efficiency, with $r_{\text{iso}}^{0.3} < 0.05$, $|d_{xy}| < 0.01$ cm, and longitudinal IP $|d_z| < 0.02$ cm.
2. Isolated, prompt electrons for charm tagger calibration in dileptonic channels: MVA ID with 80% efficiency with $r_{\text{iso}}^{0.3} < 0.15$.
3. Isolated, prompt electrons in 1L channel: MVA ID with 80% efficiency.
4. Isolated, prompt electrons in 2L channel: MVA ID with 90% efficiency.
5. Veto electrons in 0L and 1L channels: MVA ID with 90% efficiency with $p_{\text{T}} >$ 15 GeV and $r_{\text{iso}}^{0.3} < 0.1$.

The usage of various types of electron selections is elaborated in Sects. 7.2 and 9.3. The p_{T} thresholds of the reconstructed electrons are generally motivated by the p_{T} thresholds of the HLT paths corresponding to the selections they are used in.

5.3 Jets

Jets are reconstructed from stable particles found in the PF workflow that are not already associated with a prompt lepton or a photon. In particular, the anti-k_T (AK) algorithm [5, 6] is used for the reclustering process. The AK algorithm is infrared and collinear safe, which means the clustering is independent of additional soft emissions and collinear particle splittings. The jets clustered by the AK algorithm are circular in the (η, ϕ) plane around a well defined jet axis. To ensure that most particles clustered in the jet arise from the PV (and therefore are not from pileup), charged PF particles with tracks originating from any vertex other than the PV are excluded in the clustering. Such a method is referred to as Charged Hadron Subtraction (CHS) and the jets are called CHS jets. Jets used in the presented analyses are reconstructed with a distance parameter $\Delta R = 0.4$ and are usually referred to as AK4 CHS jets or "thin" jets. Additionally, a complementary analysis using a merged-jet approach that has been briefly discussed in Sect. 8.3 used AK jets with $\Delta R = 1.5$, similarly known as AK15 jets.

The jet momentum is determined as the vector sum of all particle momenta in the jet, and is found from simulation to be, on average, within 5–10% of the true momentum over the entire p_{T} spectrum and detector acceptance. Jet energy corrections are derived from simulations to bring the measured average response of jets to that of generator-level jets. In situ measurements of the momentum balance in dijet, γ + jet, Z + jet, and multijet events are used to correct any residual differences in jet energy scale (JES) in data and simulation [7, 8]. The jet energy resolution (JER) amounts typically to 15–20% at 30 GeV, 10% at 100 GeV, and 5% at 1 TeV. The resolution in simulated jets is often more optimistic than that in data, and hence additional jet energy smearing corrections are applied to simulated jets to match the resolution in data. The JES and JER corrections together are sometimes referred to as Jet Energy Correction (JECs). However, the centrally-derived JER smearing corrections are not used in this analysis, as an additional energy correction is performed for charm jet candidates (cf. Sect. 9.4.2). Instead, a dedicated smearing discussed in Sect. 9.7.3 is used.

Additional selection criteria are applied to each jet to remove jets potentially dominated by anomalous contributions (noise) from various subdetector components or reconstruction failures. The CMS JetMET POG provides so-called `JetID`s to discard jets that are potentially dominated by detector noise. The presented analyses use the so-called tight `JetID`, that applies η-dependent criteria on jet composition variables, like neutral and charged EM and hadron energy fractions, muon and electron energy fractions, number of charged and neutral constituents, etc. Furthermore, jets arising from pileup events are discarded using a `PileupJetID` [9, 10], which is also provided centrally by the CMS JetMET POG. This ID makes use of a BDT to discard pileup jets, using jet features like trajectories of tracks associated to the jets, topology of the jet shape, object multiplicity, etc. The `PileupJetID` is applied only to jets with $p_T < 50$ GeV as high-p_T pileup jets are a rare occurrence.

In addition, jets overlapping with selected prompt electrons and muons (as described in previous subsections) within a radius of $\Delta R = 0.4$ are discarded to get rid of prompt leptons also present in the jet collection. Since heavy-flavour tagging algorithms strongly rely on the track reconstruction, and jets in this thesis are subject to flavour tagging, only jets within the tracker acceptance are used throughout this thesis.

Displaced secondary vertices (SVs) from the decays of b or c hadrons are reconstructed using the inclusive secondary vertex finding algorithm [11]. This algorithm does not a *priori* assume any matching of the jets to a SV, but instead takes as input all reconstructed tracks in the event with $p_T > 0.8$ GeV, and having a distance of closest approach from PV to track, projected onto the direction of the beam axis, below 0.3 cm. After the iterative procedure of track clustering, the final set of tracks is matched to the reconstructed jets if the distance between the direction of the SV displacement (pointing from PV to SV) and the jet axis satisfies $\Delta R < 0.4$.

5.4 Missing Transverse Energy

The measured missing transverse energy (MET or p_T^{miss} or E_T^{miss}) of an event is an important quantity that gives us the information of particles that leave the detectors undetected, such as neutrinos. In searches for dark matter and other exotic particles, MET can be interpreted in the context of undetectable particles as well. In the presented analyses, MET has been used to reconstruct $Z \to \nu\nu$ decays as well as $W \to l\nu$ decays.

While the z-component (that is, along the beam axis) of the momentum of the initial parton-parton collision system can have an arbitrary value, the momentum along the transverse plane can be assumed to be negligible. Therefore, by the conservation of momentum, the vector sum of the transverse momenta of the final-state particles should equal zero. If certain particles are not reconstructed, the vector sum of their momenta can hence be expressed as the negative of the vector sum of the momenta of the reconstructed particles [12], as

$$\vec{p}_{\mathrm{T}}^{\mathrm{miss}} = - \sum_{i \in \text{measured particles}} \vec{p}_{\mathrm{T},i}. \tag{5.1}$$

The measurement of MET is dependent on the measurement of the momenta of all reconstructed particles. In particular, JES calibrations are seen to have a significant effect on MET. Therefore, whenever JECs are applied, the MET is recalculated with the updated energies of the jets. In general, a very good agreement of MET is observed between data and simulated events, and therefore no additional corrections are applied specifically for simulated MET [12].

References

1. CMS Collaboration (2012) Performance of CMS muon reconstruction in pp collision events at $\sqrt{s} = 7$ TeV. JINST 7:P10002. https://doi.org/10.1088/1748-0221/7/10/P10002. arXiv:1206.4071
2. CMS Collaboration (2018) Performance of the CMS muon detector and muon reconstruction with proton-proton collisions at $\sqrt{s} = 13$ TeV. JINST 13:P06015. https://doi.org/10.1088/1748-0221/13/06/P06015. arXiv:1804.04528
3. CMS Collaboration (2015) Performance of electron reconstruction and selection with the CMS detector in proton-proton collisions at $\sqrt{s} = 8$ TeV. JINST 10:P06005. https://doi.org/10.1088/1748-0221/10/06/P06005. arXiv:1502.02701
4. CMS Collaboration (2021) Electron and photon reconstruction and identification with the CMS experiment at the CERN LHC. JINST 16:P05014. https://doi.org/10.1088/1748-0221/16/05/p05014. arXiv:2012.06888
5. Cacciari M, Salam GP, Soyez G (2008) The anti-k_{T} jet clustering algorithm. JHEP 04:063. https://doi.org/10.1088/1126-6708/2008/04/063. arXiv:0802.1189
6. Cacciari M, Salam GP, Soyez G (2012) FastJet user manual. Eur Phys J C 72:1896. https://doi.org/10.1140/epjc/s10052-012-1896-2. arXiv:1111.6097
7. CMS Collaboration (2017) Jet energy scale and resolution in the CMS experiment in pp collisions at 8 TeV. JINST 12:P02014. https://doi.org/10.1088/1748-0221/12/02/P02014. arXiv:1607.03663
8. CMS Collaboration (2018) Jet energy scale and resolution performance with 13 TeV data collected by CMS in 2016. CMS Detector Performance Note CMS-DP-2018-028. https://cds.cern.ch/record/2622157
9. CMS Collaboration (2020) Pileup mitigation at CMS in 13 TeV data. JINST 15:P09018. https://doi.org/10.1088/1748-0221/15/09/P09018. arXiv:2003.00503

10. CMS Collaboration (2020) Performance of the pile up jet identification in CMS for Run 2. CMS Detector Performance Note CMS-DP-2020-020. https://cds.cern.ch/record/2715906
11. CMS Collaboration (2011) Measurement of $\mathrm{B\overline{B}}$ angular correlations based on secondary vertex reconstruction at $\sqrt{s} = 7$ TeV. JHEP 03:136. https://doi.org/10.1007/JHEP03(2011)136. arXiv:1102.3194
12. CMS Collaboration (2019) Performance of missing transverse momentum reconstruction in proton-proton collisions at $\sqrt{s} = 13$ TeV using the CMS detector. JINST 14:P07004. https://doi.org/10.1088/1748-0221/14/07/P07004. arXiv:1903.06078

Chapter 6
Flavour Tagging

This section discusses the flavour tagging approaches relevant to the analyses described in the thesis. It closely follows the related section in Ref. [1].

6.1 Jet Flavour Identification

The signal processes studied in the presented analyses all contain one or more charm quarks in their final states. While all energetic quarks produced from the hard scattering process in the collision, in general, form jets in the final state, jets arising from heavy flavour quarks (b and c) have certain features that distinguish them from light flavour (udsg) jets and from each other. This is because the long-lived b and c hadrons originating from the hadronisation of the respective quarks travel a distance ranging typically from $\sim$ 1 mm to $\sim$ 1 cm before further decaying at a SV. The displacement of the SV from the PV can be resolved by exploiting the high resolution of the CMS tracker. Furthermore, the reconstructed invariant mass, computed from the four-vectors of the particles that are assigned to such an SV, is strongly correlated with the mass of the b or c hadron. Therefore, it is possible to use this information to determine the flavour of the quark (or gluon) that gave rise to a jet, an approach known as flavour tagging. The existing heavy-flavour tagging algorithms exploit these properties by combining the features of tracks inside the jet (momentum, displacement, multiplicity) with the features of reconstructed SVs (mass and both the direction and the magnitude of the SV displacement from the PV).

A generator-level definition of the jet flavour in simulated jets is needed to train the heavy-flavour tagging algorithms, as well as to calibrate them using data (as described in Sect. 7). This definition is based on a procedure referred to as ghost-matching [2], which involves reclustering the generator-level jet constituents after adding the intermediately decayed b or c hadrons to the list of particles used for the clustering. In case there are multiple b (c) hadrons in a decay chain, only the last b (c) hadron in the chain, i.e., the b (c) hadron that does not further decay into another b (c)

S. Mondal, *Charming Decays of the Higgs, Z, and W Bosons*, Springer Theses,
https://doi.org/10.1007/978-3-031-84677-9_6

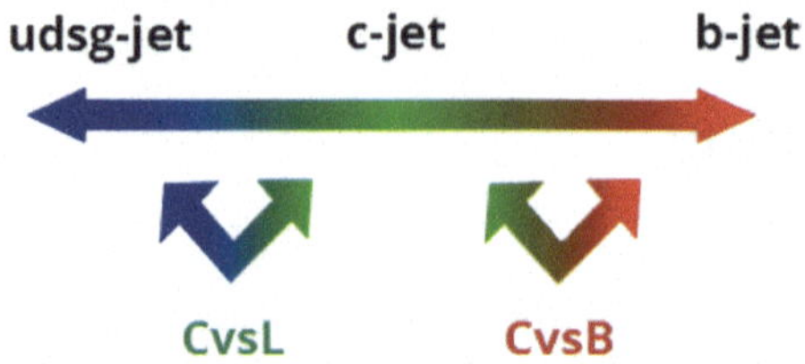

Fig. 6.1 An illustration of the properties of udsg jets, c jets, and b jets. The distribution of any jet variable for c jets is intermediate to that for udsg and b jets. Therefore, two different discriminators, CvsL and CvsB are defined to discriminate c from light-flavour jets and b jets, respectively

hadron, is reclustered. The moduli of the four-momenta of these generated hadrons are rescaled to a very small number (resulting in so-called ghost hadrons), to ensure that they do not affect the reconstructed jet momentum, and only their directional information is kept. If at least one ghost b hadron has been clustered inside the jet, it is referred to as a bottom jet. If no b hadron is found, but instead at least one c ghost hadron is clustered inside the jet, it is referred to as a charm jet. In all other cases the jet is labelled as a light-flavour jet, and hence this label includes jets originating from light quarks (uds), as well as gluon-initiated (g) jets. Particles from pileup interactions are not included in the reclustering.

Since the average lifetime and mass of b hadrons is larger than that of c hadrons, it is easier to distinguish signatures of b jets from those of light-flavour jets, than it is to distinguish those of c jets from those of light-flavour jets. The fact that the properties of c jets are distributed midway between those of light-flavour and b jets requires the definition of two discriminating variables; one to distinguish c jets from light-flavour jets (CvsL) and one to distinguish c jets from b jets (CvsB), as illustrated in Fig. 6.1. Historically, the first c jet identification algorithm used in the CMS Collaboration was based on a combination of two BDTs, each trained for one of the above mentioned discrimination purposes [3]. The state-of-the-art heavy-flavour tagging algorithms are currently based on multi-class DNN architectures and predict separate probabilities for each jet to originate from a given quark flavour (or gluon). Through an appropriate combination of the probability values computed by these algorithms, these taggers can function either as a bottom jet or as a charm jet identification algorithm.

6.2 Flavour Tagging Algorithms

One of the baseline taggers used in CMS is called DeepCSV [4]. The DNN architecture of the DeepCSV algorithm is composed of five fully connected hidden layers with 100 nodes in each layer. It takes as input a set of 66 reconstructed observables related to the charged-particle tracks and secondary vertices that are assigned to a given jet, and outputs four probabilities, P(b), P(bb), P(c) and P(udsg), that denote

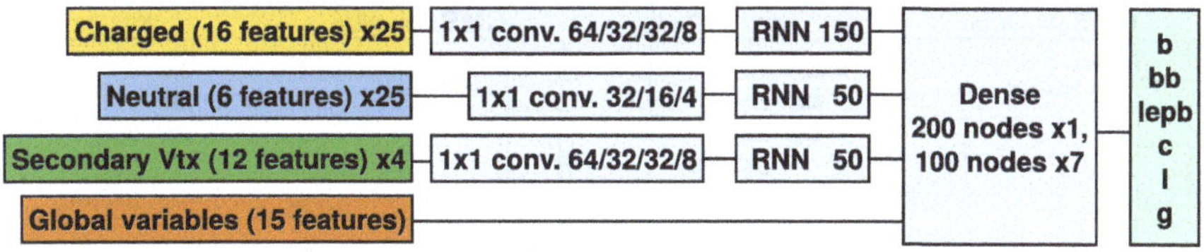

Fig. 6.2 An illustration of the DeepJet architecture. Three separate branches are used to process charged candidates, neutral candidates and secondary vertices. The algorithm makes use of 1x1 convolutional layers to perform automatic feature engineering for each class of jet constituents. The three RNN (LSTM) layers combine the information for each sequence of constituents. Finally the full jet information is combined using fully connected layers [5]

the probability of a jet to originate from one b quark, two b quarks merged into the same jet, one or more c quarks, or a light-flavour quark (or gluon), respectively.

The tagging algorithm used for the presented analyses is known as DeepJet [5, 6]. This algorithm uses an architecture composed of subsequent convolutional, recurrent and fully connected hidden layers. Its input is composed of a set of up to 613 observables related to charged and neutral PF candidates (without a *priori* selection criteria and without explicitly classifying charged PF candidates as charged hadron or leptons, and neutral PF candidates as photons or neutral hadrons) as well as the SVs that are assigned to the jet. Apart from the fact that DeepJet exhibits a higher-dimensional input space and a more complex architecture, it further subdivides the output classes into additional categories. In addition to the DeepCSV output categories, P(b_{lep}) is added to identify leptonic b hadron decays and P(udsg) is split further into P(uds) and P(g) with the goal of separately identifying jets originating from light quarks and gluons, respectively. More detailed information on the inputs, architecture, and training of these algorithms can be found in Refs. [4–6]. An illustration of the DeepJet architecture is presented in Fig. 6.2.

These output probabilities can be appropriately combined to define a set of b and c tagging discriminators as summarised in Table 6.1. For b jet identification, a discriminant is defined to distinguish b jets from either c or light-flavour jets using one single discriminator (BvsC/L). For c jet identification, two distinct discriminators are defined as the ratios in the second and third columns in Table 6.1. The normalised distributions of these discriminators for both algorithms are shown in Fig. 6.3 for jets with $p_T > 20\,\text{GeV}$ lying within the η range of the tracker, taken from hadronically

Table 6.1 Summary of the heavy-flavour tagging definitions for both b and c tagging using the DeepCSV and DeepJet taggers. P(a) represents the probability of having an "a"-type jet (see text)

Tagger	BvsC/L	CvsB	CvsL
DeepCSV	$P(\text{b})+P(\text{bb})$	$\frac{P(\text{c})}{P(\text{c})+P(\text{b})+P(\text{bb})}$	$\frac{P(\text{c})}{P(\text{c})+P(\text{udsg})}$
DeepJet	$P(\text{b})+P(\text{bb})+P(\text{b}_{\text{lep}})$	$\frac{P(\text{c})}{P(\text{c})+P(\text{b})+P(\text{bb})+P(\text{b}_{\text{lep}})}$	$\frac{P(\text{c})}{P(\text{c})+P(\text{uds})+P(\text{g})}$

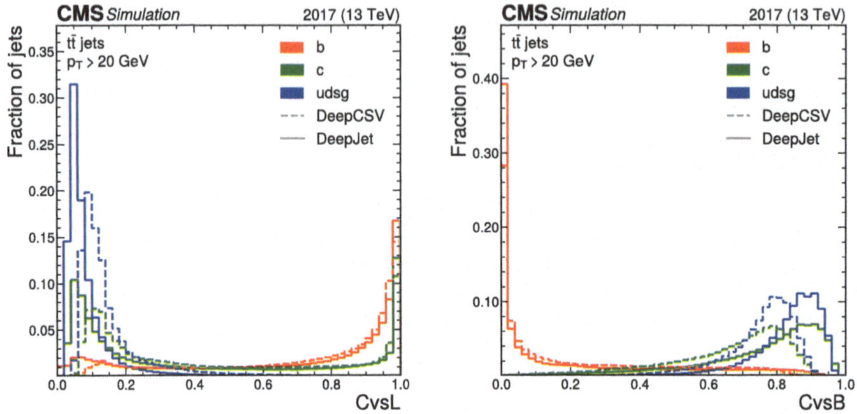

Fig. 6.3 Unit-normalised distributions of the CvsL (left) and CvsB (right) discriminators for the DeepCSV (dashed) and DeepJet (solid) algorithms using jets from simulated hadronic $t\bar{t}$ events with $p_T > 20$ GeV and $|\eta| < 2.5$. The distributions are shown for b (red), c (green) and light-flavour jets (blue) separately

decaying $t\bar{t}$ events simulated using 2017 data-taking conditions. The performance of these algorithms can be assessed by evaluating the selection efficiency for c jets as a function of the misidentification rate for either b or light-flavour jets for different selection thresholds on the discriminator values, giving rise to so-called ROC curves, as explained in Sect. 4.7.1. The ROC curves for samples simulated using 2017 data-taking conditions are shown in Fig. 6.4. It can be seen that the DeepJet algorithm has a lower mistagging efficiency than the DeepCSV algorithm in both CvsL and CvsB discrimination. Furthermore, the selection efficiency for c jets can be evaluated as a simultaneous function of b and light-flavour jet misidentification rates, which produces a two-dimensional (2D) ROC contour plot as shown in Fig. 6.5. This plot shows that the DeepJet algorithm outperforms the DeepCSV algorithm in simultaneous CvsL and CvsB discrimination over the entire 2D phase space, as well. The DeepCSV algorithm itself has already shown a significantly improved performance over the original c tagging algorithm, which was based on a combination of two BDTs, demonstrating the significant advancements that have been made in heavy-flavour identification over the last six years.

The 2D (normalised) distributions of the CvsL and CvsB discriminators of both algorithms are shown in Fig. 6.6 for different jet flavours for jets in samples simulated using 2017 data-taking conditions. In this 2D phase space spanned by the CvsL and CvsB discriminators, light-flavour jets are situated almost exclusively in the upper left corner, whereas c jets have a significant fraction along the right edge, and b jets are distributed largely towards the lower right corner, as expected.

An a *priori* track selection is applied to define the collection of tracks considered as input to the DeepCSV algorithm. This selection is optimised to identify tracks from b and c hadrons, while rejecting tracks from pileup interactions and poorly reconstructed tracks that do not match any genuine particle passing through the detector.

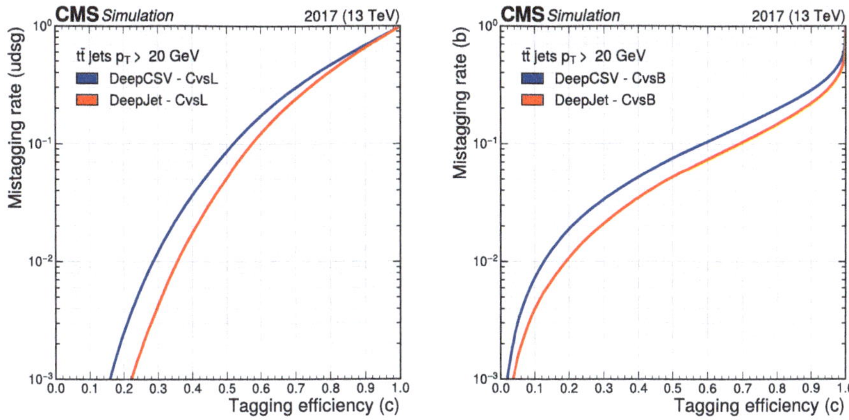

Fig. 6.4 The ROC curves showing the individual performance of the CvsL (left) and CvsB (right) discriminators for the DeepCSV (blue) and DeepJet (red) algorithms using jets from simulated hadronic tt̄ events with $p_{\mathrm{T}} > 20\,\mathrm{GeV}$ and $|\eta| < 2.5$ under the 2017 data-taking conditions

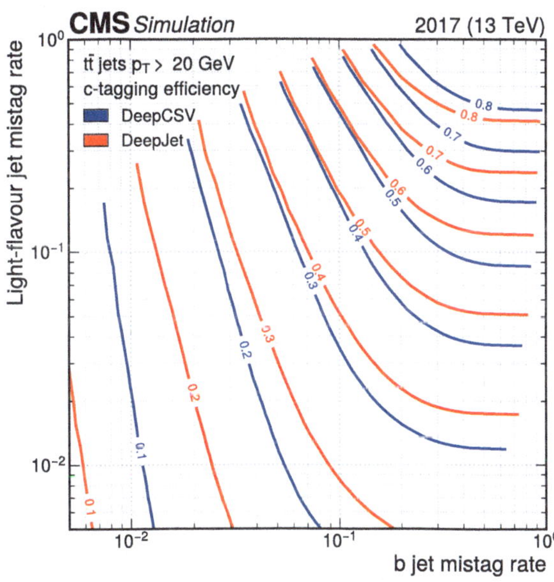

Fig. 6.5 Two-dimensional ROC contours showing the c tagging efficiency as simultaneous functions of b jet and light-flavour jet mistagging rates for DeepCSV (blue lines) and DeepJet (red lines) algorithms using jets with $p_{\mathrm{T}} > 20\,\mathrm{GeV}$ and $|\eta| < 2.5$, from simulated hadronically decaying tt̄ events under the 2017 data-taking conditions. Each line represents points in the plane that correspond to a fixed value of the c tagging efficiency, which is shown as a number at the centre of each line

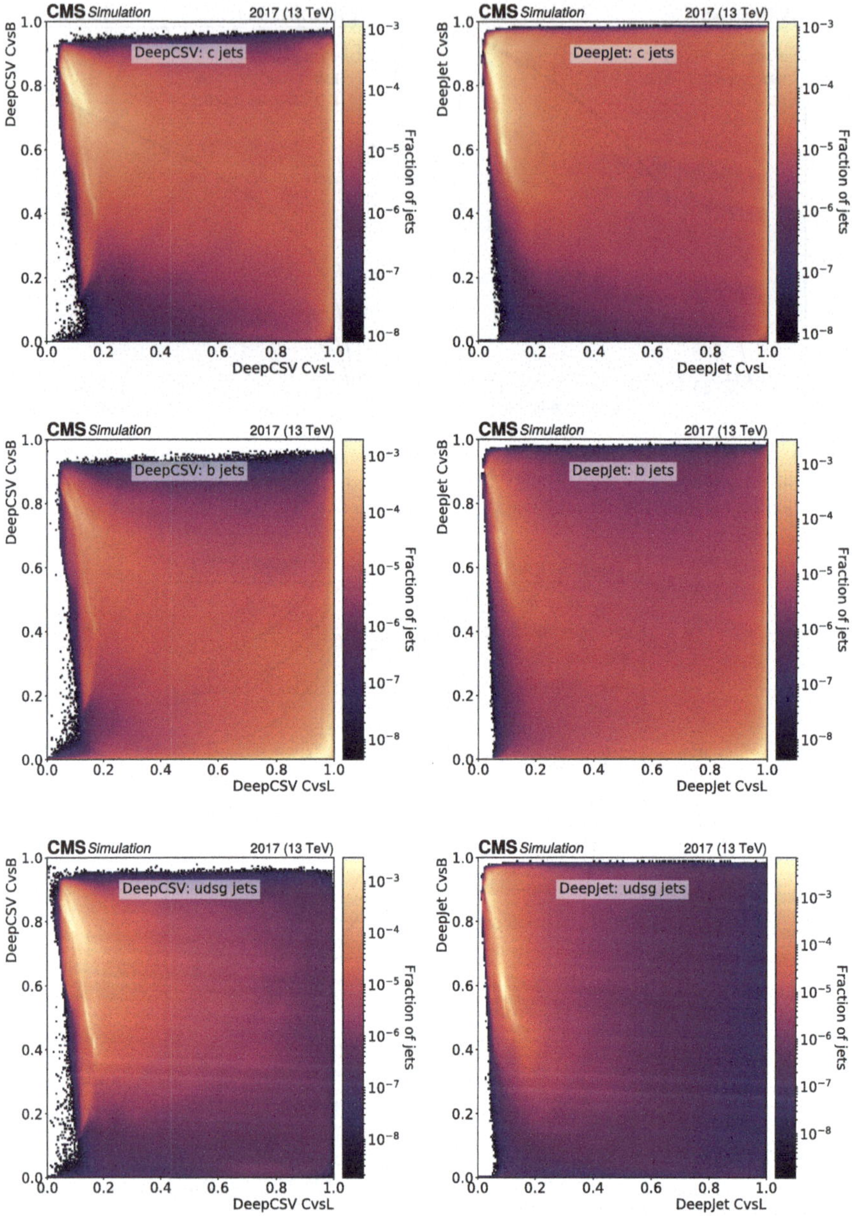

Fig. 6.6 Normalised 2D distributions showing the CvsL and CvsB discriminators on the x and y axes, respectively. Distributions are shown using c (upper), b (middle) and light-flavour (lower) jets with $p_T > 20\,\text{GeV}$ and $|\eta| < 2.5$ from simulated hadronically decaying $t\bar{t}$ events using the dataset corresponding to the 2017 data-taking conditions. The left-hand column shows the discriminators of the DeepCSV algorithm, whereas the right-hand column shows those of the DeepJet algorithm

It is therefore possible that a given jet has no tracks that pass this preselection and therefore no information is present to calculate a heavy flavour tagging discriminant. These jets are assigned a default output value of -1. Given the known differences in track reconstruction efficiency and the poorly reconstructed track rate between the simulation and collision data, it is important to calibrate the rate of jets with such default discriminator values.

No such preselection criteria are used in the DeepJet algorithm, which instead takes as input the entire collection of PF candidates associated with a given jet. Therefore, no default values are expected to appear in the output probabilities of the DeepJet algorithm. Nevertheless, because of the definition of the CvsL and CvsB discriminators shown in Table 6.1, an undefined discriminator value can appear if the denominator becomes zero. This situation is observed when the b jet probability, P(b)+P(bb)+P(b_{lep}), is evaluated to be exactly 1 by the DeepJet algorithm, resulting by construction in an output value of 0 for all other output probabilities. Consequently, this appears almost exclusively for b jets. Whenever such a situation appears for a given jet, its c tagging discriminator values, CvsL and CvsB, are evaluated to be undefined and 0, respectively, and hence constitute a special category of jets separated from the continuous jet distribution in the 2D CvsL–CvsB plane. For simpler representation, both CvsL and CvsB values of these jets are defaulted to a value of -1 in this thesis.

6.3 Mismodelling in Simulated c Tagging Distributions

The need for calibrating the heavy-flavour tagging discriminants arises from possible mismodelling of the simulated inputs to the DNNs. Important discriminating properties such as the track displacement or the SV displacement are not very well modelled in simulation and are subject to changes in the detector alignment. As already evident from other heavy-flavour calibration methods [4], the performance of the heavy-flavour tagging algorithms is overestimated in the simulation, resulting in higher simulated b and c tagging efficiencies and lower misidentification probabilities compared with those observed in the data. Rather than correcting the efficiency of a selection in the discriminator, the full simulated differential shape of that discriminator can be calibrated to match the shape observed in data. Such a strategy has been developed in the past for the b tagging discriminator distribution [4]. For the first time, Ref. [1] presented a method of calibrating the differential c tagging discriminator shapes, and the same method has been reported in this thesis in Sect. 7. The results from this method have then been implemented in the analyses described in the subsequent sections. Since c jet identification relies on both the CvsL and CvsB discriminators simultaneously, the calibration of the c tagger is performed as functions of both the CvsL and CvsB distributions. Jets with defaulted values of CvsL and CvsB are calibrated as a separate category. Section 7 is devoted to the discussion of this novel calibration method.

References

1. CMS Collaboration (2022) A new calibration method for charm jet identification validated with proton-proton collision events at $\sqrt{s}$ =13 TeV. JINST 17:P03014. https://doi.org/10.1088/1748-0221/17/03/P03014. arXiv:2111.03027
2. Cacciari M, Salam GP (2008) Pileup subtraction using jet areas. Phys Lett B 659:119. https://doi.org/10.1016/j.physletb.2007.09.077. arXiv:0707.1378
3. CMS Collaboration (2016) Identification of c-quark jets at the CMS experiment. CMS Physics Analysis Summary CMS-PAS-BTV-16-001. http://cds.cern.ch/record/2205149
4. CMS Collaboration (2018) Identification of heavy-flavour jets with the CMS detector in pp collisions at 13 TeV. JINST 13:P05011. https://doi.org/10.1088/1748-0221/13/05/P05011. arXiv:1712.07158
5. Bols E et al (2020) Jet flavour classification using DeepJet. JINST 15:P12012. https://doi.org/10.1088/1748-0221/15/12/P12012. arXiv:2008.10519
6. CMS Collaboration (2018) Performance of the DeepJet b tagging algorithm using 41.9/fb of data from proton-proton collisions at 13 TeV with Phase 1 CMS detector. CMS Detector Performance Note CMS-DP-2018-058. http://cds.cern.ch/record/2646773

Chapter 7
Charm Tagger Calibration

This section details the novel method used to calibrate the full differential shapes of the charm tagger distributions. The method and the results were first published in Ref. [1] and this section of the thesis closely follows relevant sections of the reference.

The general idea of the calibration method is to define three selections of jets in data that are enriched in c, b, and light-flavour jets respectively. The same selections when applied on simulated events, lets us compare the agreement of the discriminator shapes between data and simulation. Following this, an iterative fit method is implemented to alter the simulated jet yields in fine bins in the CvsL–CvsB 2D plane, with an aim to minimize a χ^2-based metric, that quantifies the difference in the tagger distributions between data and simulation. Finally, the results are presented as "scale factors" (SFs), that can be applied as weights to simulated jets to mimic the tagger performance in data.

7.1 Data and Simulated Samples

The data used to derive the calibration of the c tagger discriminator distributions were collected from pp collisions recorded at a centre-of-mass energy of 13 TeV with the CMS detector in the Run-2 of the LHC. Calibration factors are derived separately for datasets recorded in the years 2016, 2017, and 2018. The integrated luminosities per year are around 35.9, 41.5 and 59.9 fb^{-1}, respectively.

The data is selected using a set of single-lepton and dilepton triggers. The dielectron (dimuon) trigger requires at least two isolated electrons (muons), one with $p_T >$ 23 (17) GeV and another with $p_T >$ 12 (8) GeV. The electron-muon trigger requires the presence of at least one electron and at least one muon, where the lepton with the largest p_T is required to have $p_T >$ 23 GeV and the one with the second-largest p_T to have $p_T >$ 12 (8) GeV if it is an electron (muon). Finally, for the single-lepton

S. Mondal, *Charming Decays of the Higgs, Z, and W Bosons*, Springer Theses,
https://doi.org/10.1007/978-3-031-84677-9_7

triggers, the thresholds are dependent on the year of data collection and the HLT paths deployed in the respective years:

2016: The single-electron (single-muon) trigger requires the presence of at least one isolated electron (muon) with a p_T above 27 (24) GeV.
2017: The single-electron (single-muon) trigger requires the presence of at least one isolated electron (muon) with a p_T above 32 (27) GeV.
2018: The single-electron (single-muon) trigger requires the presence of at least one isolated electron (muon) with a p_T above 32 (24) GeV.

The simulated samples were generated using standard MC techniques as outlined in Sect. 4.2. A brief description for each sample is presented as follows:

- The ME generation of the $t\bar{t}$ simulation is performed with POWHEG (V2) at NLO accuracy in QCD, and its cross section is scaled to a theoretical prediction at next-to-NLO (NNLO) in QCD, including resummation of next-to-next-to-leading logarithmic soft-gluon terms.
- The ME generation of the W+jets and DY processes is performed with MADGRAPH5_AMC@NLO at LO precision, with MLM jet matching and the cross sections normalised to NNLO calculations [2].
- The single top quark production in the s-channel was also simulated using MADGRAPH5_AMC@NLO in the four-flavour scheme (four lightest quarks as the initial-state partons), whereas the single top quark production in the t-channel was simulated using POWHEG also in the four-flavour scheme. The tW channel was simulated using POWHEG in the five-flavour scheme [3, 4], with its cross section normalised to the NLO calculations [5].
- The diboson samples are simulated at LO using PYTHIA for the ME generation. Their cross sections are normalised to those calculated at NLO for WZ and ZZ [6], and NNLO for WW [7].

The simulated physics processes used are dependent on the selection being considered (selections described in Sect. 7.2), and are listed in Table 7.1 along with the respective theoretical cross sections that the processes were scaled to using Eq. 4.1.

7.2 Event Selection

The calibration strategy involves identifying three samples of jets in the data that are enriched respectively in c, b, and light-flavour jets. This has resulted in the definition of three jet samples enriched either in W+c events (c enriched), $t\bar{t}$ events (b enriched), and DY+jet events (udsg enriched). The flavour tagging algorithms being calibrated are not used in the construction of these selections to keep the resulting calibration free from potential biases. However, the sampled c and b jets are required to contain a soft muon inside the jet cone, to enrich the samples with jets containing semileptonically decaying c and b hadrons, which increases signal purity. Since soft muons within

Table 7.1 The simulated samples with their cross sections (XSec) used in the various selections for charm tagger calibration. "Incl." stands for inclusive, "Dilep." stands for dileptonic, and "semi" and "Semilep." stand for semileptonic. MG5 refers to MADGRAPH5_AMC@NLO used in LO mode, while AMC@NLO refers to the same used in NLO mode. Whenever the inclusive W+jets sample is used alongside jet multiplicity-split W+jets samples, a generator level requirement of # jets < 1 is applied on the inclusive samples. Inclusive decays of $t\bar{t}$ have been used for $t\bar{t}$ samples corresponding to the 2016 data-taking conditions, while those with the 2017 and 2018 data-taking conditions were split depending on the decay mode

Process	Samples	Generator	Used in selection?			XSec (pb)
			W+c/semi $t\bar{t}$	DY+jet	dilep $t\bar{t}$	
W+Jets	W($l\nu$)+Jets (incl.)	MG5	✓		✓	61526
	W($l\nu$)+1 jet		✓			11786
	W($l\nu$)+2 jet		✓			3854
	W($l\nu$)+3 jet		✓			1274
	W($l\nu$)+4 jet		✓			701
DY+Jets	$m_{ll} > 50$ GeV	MG5	✓	✓	✓	6225
	$10 < m_{ll} < 50$ (GeV)				✓	18610
$t\bar{t}$	Incl. $t\bar{t}$ (2016)	POWHEG	✓		✓	731
	Dilep. $t\bar{t}$		✓		✓	88
	Semilep. $t\bar{t}$		✓		✓	365
	Hadronic $t\bar{t}$		✓		✓	378
SingleTop	s-channel	AMC@NLO	✓		✓	10
	t-channel t	POWHEG	✓		✓	44
	t-channel $\bar{t}$		✓		✓	26
	tW		✓		✓	36
	$\bar{t}$W		✓		✓	36
Diboson	WW	PYTHIA	✓		✓	64
	WZ		✓		✓	23
	ZZ		✓		✓	10

jets are treated as any other charged PF candidate in both DeepCSV and DeepJet trainings, as opposed to training the network with explicit soft muon information, the bias in the discriminator responses arising from using muon-containing c and b jets as a proxy for all c and b jets, is expected to be minimal. In Sect. 7.8.2, applicability of the correction factors to inclusive samples despite the nonuniversality of the c and b jets used for calibration is demonstrated.

The SM physics processes, namely, W+c, $t\bar{t}$ and DY+jet, chosen to yield the three jet samples, were experimentally studied with great precision by the CMS experiment [8–11]. These analyses did not observe any significant deviation from the SM expectations. Therefore, effects of BSM physics, if any, on these samples are expected to be insignificant when compared with the statistical and systematic uncertainties associated with the data and simulation. Hence, the distributions of the c tagger outputs for these jet samples in data are expected to agree with the corresponding calibrated distributions in simulation within uncertainties.

In the following subsections, the event selection criteria to obtain each of these three topologies are outlined, followed by a discussion on the purity of the different jet flavours after each selection.

7.2.1 The c Jet Enriched Selection

Charm jets in data are selected from events with a jet produced in association with a W boson, following the strategy discussed in the CMS W+c cross section analysis [8] and the c jet identification SF derivation [12]. The relevant events involve a leptonically decaying W boson (W $\rightarrow l\nu$, $l = e, \mu$) and a jet originating from the hadronisation of a charm quark. These charm jets are identified using the semileptonic decay of the c hadron, which produces a soft muon within the jet in the final state. Charm hadron decays into electrons are not considered, because of the lower efficiency of electron reconstruction at low energies [13]. Thus, the targeted final state consists of an isolated charged lepton (electron or muon) from the W boson decay, MET due to the presence of a neutrino from the W boson decay, and at least one jet with a soft, nonisolated muon inside it. The selected events are divided into two categories:

- *Opposite-sign (OS) events*, where the soft muon inside the jet and the isolated lepton from the W boson have opposite charges;
- *Same-sign (SS) events*, where the soft muon inside the jet and the isolated lepton from the W boson have the same charge.

If there are multiple soft muons inside the jet, the muon with the highest p_T determines the sign of the charge. The major background for the W+c selection, as described above, is the production of a W boson associated with the radiation of a virtual gluon that decays into a pair of b or c quarks, with one or more of the resulting jets in the final state having a soft muon inside it. These backgrounds have 50%–50% probabilities of being tagged as OS or SS. In contrast, the leading-order contributions to the production of a W boson with one c jet always result in an OS signature, as illustrated in Fig. 7.1. Hence, an OS-SS subtraction largely reduces the major background and yields a distribution enriched in W+c events. All distributions related to W+c selection in this section are OS-SS subtracted. A few details of the selection are described as follows:

Isolated lepton: A charged lepton (electron or muon) from the W boson decay satisfying the "isolated, prompt leptons for charm tagger calibration" criteria explained in Sects. 5.1 and 5.2 is required. The events are categorised into electron and muon channels, depending on the flavour of each charged lepton. The electron channel requires exactly one isolated electron with a p_T of at least 30 (34) GeV

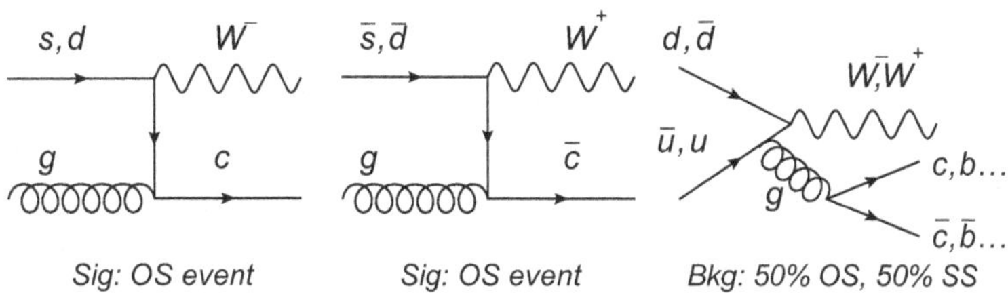

Fig. 7.1 Feynman diagrams showing production of charm quarks in association with a W boson (left and middle) and the major background (right)

and no isolated muon, and the muon channel requires exactly one isolated muon with a p_T of at least 27 (30) GeV and no isolated electron, for the data-taking year 2016 (2017 and 2018). The events are triggered by a single-electron and/or single-muon HLT. The W bosons decaying into τ leptons are not explicitly excluded, and may enter the selection through their leptonic decay products.

W boson candidate: The reconstructed W boson transverse mass is defined as

$$m_T = \sqrt{2p_T^{\text{lep}}\,\vec{p}_T^{\text{miss}}\left(1 - \cos\left(\phi_{\text{lep}} - \phi_{\vec{p}_T^{\text{miss}}}\right)\right)}, \tag{7.1}$$

where p_T^{lep} is the p_T of the isolated lepton and ϕ_{lep} ($\phi_{\vec{p}_T^{\text{miss}}}$) is the ϕ of the isolated lepton (missing momentum). An m_T of at least 50 GeV is required to ensure the presence of a neutrino produced from a W boson decay.

Jets: Events are required to have at least one and at most three jets lying inside the tracker coverage and with $p_T > 20$ GeV. At least one of these jets is required to contain a soft muon satisfying the criteria of "non-isolated, soft muons for charm tagger calibration", outlined in Sect. 5.1, with a relative isolation of at least 0.2 (0.5) in the electron (muon) channel. If there are multiple such jets in an event, the jet with the highest p_T (referred to as the muon jet) is selected. Only jets separated from the isolated lepton (from W boson decay) by $\Delta R > 0.5$ are considered.

DY suppression: To discard events where a prompt muon is misidentified as the muon jet, an upper threshold of 0.4 (0.6) is applied to the muon energy fraction of the muon jet in the muon (electron) channel. For only the muon channel, the sum of the muon and the neutral electromagnetic energy fractions of the muon jet is required to be smaller than 0.7. These two requirements heavily suppress the DY background in the muon channel by rejecting events where one of the prompt muons is misidentified as the muon jet, e.g. when the prompt muon undergoes final-state photon radiation. Furthermore, the invariant mass of the muon inside the jet and the isolated muon in the muon channel cannot match that of the Z boson (in between 80 and 100 GeV) or other dimuon resonances (below 12 GeV).

The muon jets selected in these events are the c jet candidates. The OS-SS distributions of the CvsL and CvsB variables of these candidates are presented in Fig. 7.2 (upper) for the DeepJet tagger. These plots, as well as all other jet

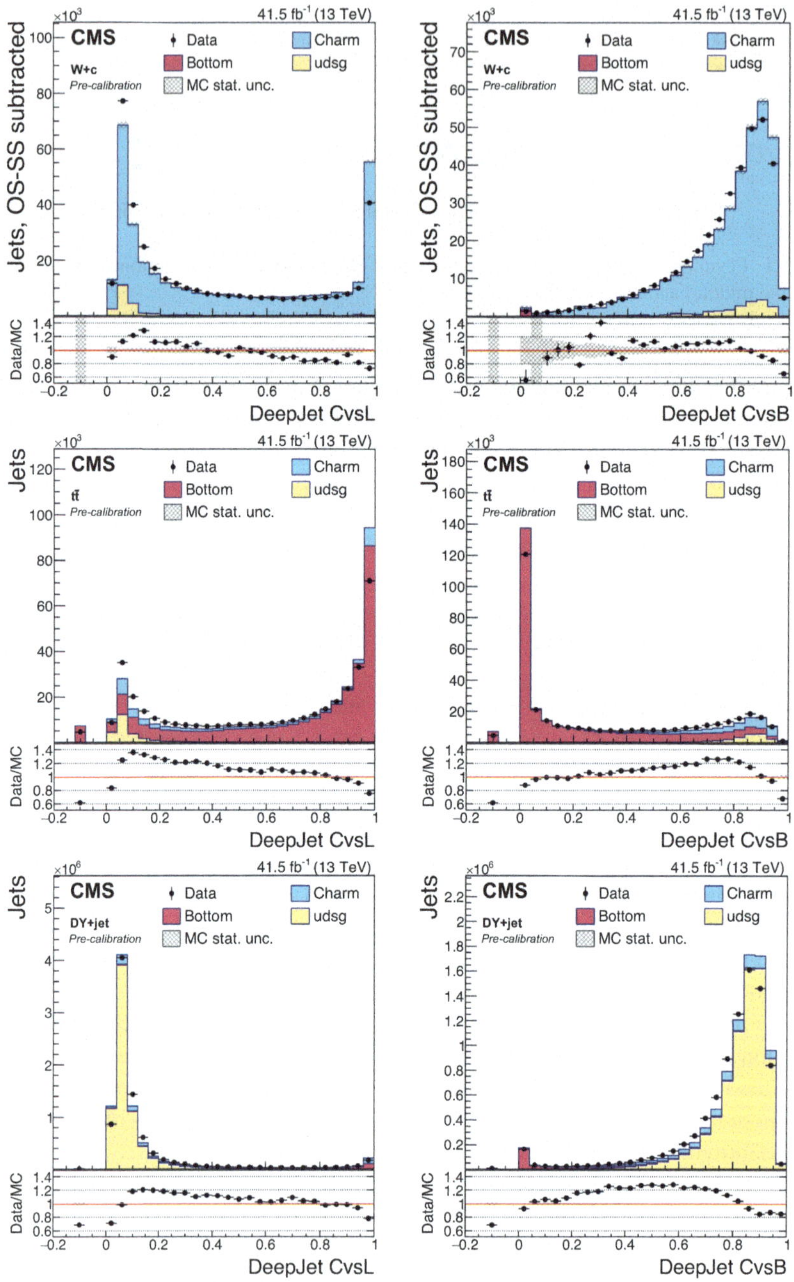

Fig. 7.2 Pre-calibration distributions of CvsL (left) and CvsB (right) obtained from the DeepJet taggers for jets selected in the W+c (OS-SS) (upper), merged semi- and dileptonic $t\bar{t}$ (middle), and DY+jet (lower) selections. The bin corresponding to a tagger value of -1 is plotted at -0.1. Vertical error bars in data represent statistical uncertainties in data. The simulations are shown as stacked histograms

distribution plots in this section, are plotted after scaling the simulated jets such that the total number of simulated jets matches the number of jets in data in every channel of every selection. This ensures that any remaining discrepancies between data and simulation, in terms of cross sections of the contributing processes, are largely reduced.

7.2.2 The B Jet Enriched Selection

The b jet enriched sample is obtained from $t\bar{t}$ events by selecting one of the b jets originating from the decay of the two top quarks. The selection includes both the semileptonic and dileptonic decay channels of the $t\bar{t}$ process, each of which is discussed separately below.

Semileptonic $t\bar{t}$ selection The strategy for the semileptonic $t\bar{t}$ selection is to select one of the b jets produced in a semileptonic $t\bar{t}$ decay by identifying a leptonically decaying W boson, a jet with a soft muon from a bottom hadron decay chain inside it, and several additional jets from the other b quark and the hadronically decaying W boson. The selection is similar to that of the W+c discussed in the previous subsection and is kept orthogonal to the latter by requiring at least four jets in the event.

Dileptonic $t\bar{t}$ selection The dileptonic $t\bar{t}$ selection chooses one of the two b jets in dileptonic $t\bar{t}$ events. The target final state consists of two isolated, oppositely charged leptons from the two W boson decays, and a minimum of two jets, at least one of which has a soft muon inside it. The selection can be subdivided into three channels depending on the flavour of the isolated leptons, namely dielectron, dimuon and electron-muon channels. The events in those channels are triggered using the dielectron, dimuon and electron-muon HLTs, respectively.

The dielectron (dimuon) channel requires exactly two electrons (muons) having p_T of at least 27 and 15 (20 and 12) GeV, and no muons (electrons). The electron-muon channel requires either one muon with $p_T^{\mu} > 25$ GeVand one electron with $p_T^{e} > 15$ GeV, or one electron with $p_T^{e} > 27$ GeV and one muon with $p_T^{\mu} > 15$ GeV. In all cases, the charged leptons are required to satisfy the criteria of "isolated, prompt leptons for charm tagger calibration in dileptonic channels" as outlined in Sects. 5.1 and 5.2.

Although the muon-electron channel thus selected is quite pure in b jets, the two other channels are contaminated by the DY background. Therefore, in the dielectron and dimuon channels, the event is required to have a MET of at least 40 GeV, and the invariant mass of the dilepton candidate is required to be incompatible with that of the Z boson (i.e. outside the range 75–105 GeV) or other dimuon resonances (below 12 GeV), to suppress the DY contamination.

The muon jet selected in each semileptonic or dileptonic $t\bar{t}$ event is the b jet candidate. In case an event has multiple muon jets, the jet with the highest p_T is selected. The CvsL and CvsB distributions of these candidates combined from both semileptonic and dileptonic decays are presented in Fig. 7.2 (middle) for the DeepJet tagger. It has been verified that the c tagger discriminator shapes of individual flavours agree quite well between semileptonic and dileptonic decay channels for both of the taggers.

7.2.3 *Light-Flavour Jet Enriched Selection*

Jets arising from light quarks (uds) and gluons (g) are selected in events containing one or more jets produced in association with a leptonically decaying Z boson (DY+jet events). The selection is split into two channels (electron and muon), depending on the flavour of the charged leptons from the Z boson. The Z bosons decaying into τ leptons are not explicitly considered, but may enter the selection through the leptonic τ decays.

The final state in the electron (muon) channel consists of exactly two isolated, oppositely charged electrons (muons) having p_T of at least 27 and 15 (20 and 12) GeV and at least one jet. Each of the two leptons is required satisfy the criteria of "isolated, prompt leptons for charm tagger calibration in dileptonic channels" as outlined in Sects. 5.1 and 5.2, and their invariant mass is required to be within 10 GeV of the Z boson mass. The events are triggered with the dielectron and dimuon HLTs, respectively. The requirement of a soft muon inside the jet is not imposed for light-flavour jets. In case there is more than one jet in the event, the jet with the highest p_T is considered as the light-flavour jet candidate. The CvsL and CvsB distributions of these candidates are presented in Fig. 7.2 (lower) for the DeepJet tagger.

7.2.4 *Summary of Selections*

To summarise, W+c events (OS-SS subtracted), $t\bar{t}$ events, and DY+jet events are used to probe the c tagger distributions of jet samples in data enriched in c jets, b jets and light-flavour jets, respectively. These selections yield a c jet purity of 92.9%, b jet purity of 81.0% and light-flavour jet purity of 86.1% in their respective samples.

7.3 Iterative Calibration

A calibration of the full discriminator shape requires deriving SFs to correct for the data-simulation discrepancy at every value of the discriminator. Since c tagging requires using both CvsL and CvsB values of a jet simultaneously, this translates into deriving data-to-simulation SFs as functions of both these parameters to simultaneously adjust the discriminator shape in the 2D plane of CvsL and CvsB. In addition, the mismodelling in simulation is expected to be different for light-flavour, c, and b jets, as the propagation and decay of various types of hadrons are modelled differently. This entails deriving different sets of SFs for different jet flavours.

The general idea of deriving SFs under this strategy is to perform a fit iteratively on three flavour-enriched jet samples allowing the three flavour components (light, c and b) in each to float freely in several bins in the 2D plane until the difference between the distribution of jets in data and simulation in all the samples is minimised. The simultaneous usage of three samples with different flavour compositions ensures the existence of a solution, while an iterative fitting approach helps in converging to a physical value. After motivating two distinct choices of binning in the 2D phase space of the CvsL and CvsB discriminators below, the algorithm employed in the iterative fit method is discussed in detail.

7.3.1 *Adaptive Binning*

Ideally, a perfect calibration would require deriving the flavour-wise correction factors in infinitesimally small areas in the CvsL–CvsB plane. However, the smaller the bin area, the lower the yield of jets in that bin, and hence the larger the statistical uncertainty in the SF. A large bin area, on the other hand, results in largely discontinuous SFs, which could be unrepresentative and could result in an incorrect calibration. To find a trade-off, an adaptive binning approach is adopted with the aim of having smaller bin areas for SFs of a given flavour in the part of the 2D plane where the yield of jets of that flavour is large. This results in a more continuous correction in high jet count regions, and ensures optimised statistical uncertainties over the entire phase space.

In practice, the binning is kept constant along one axis and made adaptive along the other in the first run, and then the choice is reversed in the second. The 2D plane of SFs for all three flavours is first divided into ten equal-width bin-slices along one axis (the "fixed axis"). The algorithm treats each of these slices independently and proceeds to determine the binning along the other axis (the "variable" axis) on a per-bin-slice basis and independently for each flavour. Thus, for each bin slice of the SF map of a given flavour, the binning along the variable axis is sequentially increased starting from a minimum threshold, b_{min}, in steps of b_{min}, until at least one of the following two conditions is fulfilled for each bin:

- the resulting SF for that flavour in that bin has a relative statistical uncertainty smaller than ϵ_{max}, or,
- the bin width along the variable axis is equal to b_{max}.

A dedicated scan of these three parameters (b_{min}, b_{max}, and ϵ_{max}) is performed and values of $b_{\text{min}} = 0.02$, $b_{\text{max}} = 0.10$, and $\epsilon_{\text{max}} = 2\%$ are chosen as an optimal compromise between the granularity of the binning and the statistical uncertainties. An illustrative diagram of the resulting SF binning in the 2D plane is given in Fig. 7.3.

After the SF results are derived using such a binning scheme, the choice of fixed and variable axes is reversed to yield another binning scheme. For example, at first,

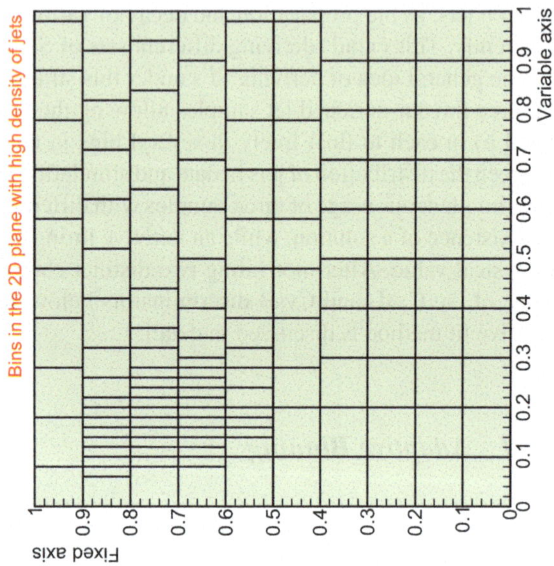

Fig. 7.3 An illustration showing the SF bin edges for a given flavour in the 2D plane as a consequence of the adaptive binning scheme. The area inside the dashed red rectangle corresponds to parts of the 2D plane containing relatively higher densities of jets of the given flavour, compared to parts outside the rectangle. The x- and y-axes represent the variable and fixed axes, respectively, so that the minimum and maximum bin widths along the x-axis are 0.02 and 0.10, respectively, while the bin width along the y-axis is fixed at 0.10

the CvsB axis could be chosen as the fixed axis and CvsL as the variable axis to yield a binning scheme and perform the iterative fit. Next, the CvsL axis could be chosen as the fixed axis and the CvsB axis as the variable one. This would yield a second binning scheme using which the iterative fit can be performed.

7.3.2 *Iterative Fit*

The iterative fit method minimises a global χ^2 value that quantifies the data-to-simulation discrepancy in the projected CvsL distribution over a given (fixed-width) CvsB bin slice, or vice versa. The method proceeds by iteratively evaluating and applying SFs for all three jet flavours in the corresponding selected topology in which they are enriched.

The algorithm begins by scaling the total number of simulated jets across all contributing processes in each channel (i.e. muon, electron, or electron-muon, as applicable) of each sample (i.e. c, b or light-flavour jet enriched) to match the number of jets selected in data in that channel. This removes any remaining data-simulation discrepancies in terms of the total jet yield, without altering the discriminator shape and simultaneously ensures that the SFs thus derived correct only the discriminator shape without significantly altering the jet yield.

Then for each fixed bin slice:

- The three samples are ranked by the purity of the samples for the given bin slice. The purity of a sample is defined as the percentage contribution of the jet flavour that the sample is meant to enrich, e.g. fraction of c jets in the W+c selection.
- Starting with the purest sample, "signal" is defined as the CvsL (CvsB) distribution of simulated jets of the flavour f_1 that this sample aims to enrich, when CvsB (CvsL) is chosen as the fixed axis. Similarly, "background" is defined as the sum of the templates of the other two flavour components of the same sample in simulation. "Signal in data" is defined as the difference between the simulated background template and the data distribution. The ratio of the "signal in data" to "signal", evaluated as a function of the quantity on the variable axis, gives the SF for f_1 in this bin slice in this iteration.[1] The ratio (an array of SFs) is calculated in discrete intervals following the binning scheme determined as explained in Sect. 7.3.1. If the yield of simulated jets of flavour f_1 in the enriched sample in any of these bins is less than one, or if the purity of the sample in any bin is less than a threshold (10%), the resulting ratio in this bin may have an unphysical central value with a large statistical uncertainty. In all such bins, a default value of 1 with a 100% statistical uncertainty is assigned as the SF.
- Moving to the second purest sample for this bin slice, the SFs for f_1 are applied to the simulated template of f_1 in this sample with the same discrete binning used to determine the SFs. Again, "signal" is defined as the template of the flavour f_2 that this distribution enriches and thus the array of SFs for f_2 is evaluated in the same way. The binning of these SFs can, in general, be different from that of f_1, and will depend on the distribution of jets of flavour f_2 in this bin slice.
- Proceeding to the third sample, the SFs for f_1 and f_2 are applied to the respective simulated templates following their respective binnings, and the SFs for f_3 corresponding to this sample are evaluated. This completes one full iteration for this bin slice.
- The next iteration begins by applying all three SF arrays to the respective simulated templates of the first sample. The SFs for f_1 are then updated. Proceeding to the next samples, f_2 and f_3 are updated likewise.
- The iterations are continued until a global χ^2 is minimised and no longer improves with further iterations. The χ^2 for a given sample, "sel" (sel = Wc, $t\bar{t}$, DY), in this bin slice is defined as

$$\chi^2_{\text{sel}} = \sum_{i=1}^{n} \frac{\left(N_{\text{sim},i} - N_{\text{data},i}\right)^2}{\sigma^2_{\text{sim},i} + \sigma^2_{\text{data},i}}, \tag{7.2}$$

[1] In practice, an array of SFs is initialised with the value 1 for all bins and all flavours with the goal of updating them to the current best estimate of the SF values after every ratio evaluation. When the difference between the current and previous best estimates after a ratio evaluation in a given bin is larger than 0.005, a change of only 0.005 is allowed in that bin. This parameter acts as a "convergence speed", and ensures a smooth convergence of the χ^2 parameter and prevents convergence to a local minimum.

where n is the number of bins of width b_{min} in the distribution along the variable axis, $N_{\text{sim},i} = \sum_{f=\text{c,b,udsg}} \text{SF}_{f,i} N_{f,i}$ is the SF-adjusted simulated jet count in the i-th bin along the variable axis, $N_{\text{data},i}$ is the jet count in data in the i-th bin, with $\sigma_{\text{sim},i}$ and $\sigma_{\text{data},i}$ as the corresponding statistical uncertainties in simulation and data, respectively.[2] The global χ^2 of a bin slice is defined as the sum of the three χ^2_{sel} values corresponding to the three samples in this bin slice, and is updated after every iteration.

An illustration of the steps involved in a single iteration for a bin slice is given in Fig. 7.4.

The same procedure is used for every bin slice for the DeepJet tagger outputs. Thus, three 2D maps of SFs corresponding to three different flavours are obtained. Next, the entire procedure is repeated with the fixed and variable axes interchanged. This gives an additional set of SFs for the same phase space, but with a different binning. The method of combining the two sets of results are presented in Sect. 7.4.

7.3.3 *Treatment of Jets with Default Discriminator Values*

Jets that are assigned a default value of -1, because they either do not have sufficient information to be processed by the tagging algorithm (DeepCSV) or have an undefined value for at least one of the c tagging discriminators (DeepJet), need to be treated separately in the calibration. Although DeepCSV is not used in the main results discussed in this thesis, the following paragraph also discusses the treatment in case of DeepCSV, for completeness.

DeepCSV Jets that have no tracks passing the DeepCSV preselection criteria, and hence have been assigned default discriminator values CvsL = CvsB = -1, remain excluded from the iterative fit treatment described in Sect. 7.3.2. Therefore, the iterative fit procedure is repeated in the same way for the jets with defaulted discriminator values in the three samples. However, only the jets from the electron channel of the W+c selection are used as the c jet enriched sample in this particular fit, because such jet candidates in the muon channel of W+c are heavily contaminated with prompt muons from DY events misidentified as jets and hence are unsuitable for use in a three-flavour fit. The χ^2 for a sample in this case is defined as:

$$\chi^2_{\text{sel}} = \frac{(N_{\text{sim}} - N_{\text{data}})^2}{\sigma^2_{\text{sim}} + \sigma^2_{\text{data}}}, \tag{7.3}$$

and global χ^2, defined as the sum of the χ^2_{sel} values of the three samples, is minimised iteratively. This yields three correction factors corresponding to the three flavours. These SFs for the bin containing jets with default DeepCSV values, when combined

[2] This quantity is always well defined since the width of any given bin is a multiple of b_{min} by construction, as explained in Sect. 7.3.1.

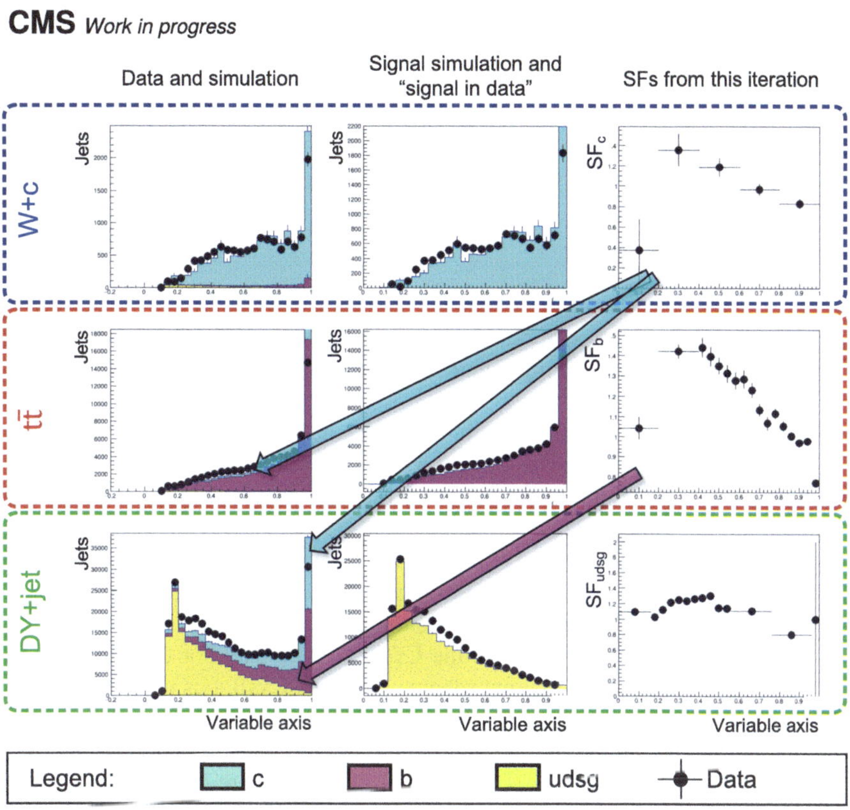

Fig. 7.4 An illustration showing the steps involved in one iteration of the iterative fit for a given bin slice. Each row corresponds to one of the three selections, sorted by signal purity (top row having highest purity). The first column shows the data-simulation comparison as a function of the variable on the variable axis, in the given bin slice. The second column in each row shows the simulated signal (defined as the flavour enriched by the selection), and its comparison with the "signal in data" (background-subtracted data). The third column shows the ratio of the signal in data to the simulated signal, and represents the best estimate of the SF distribution of each flavour in this iteration (see footnote 1) in the given bin slice. The SF binning is determined using the adaptive binning scheme (cf. Sect. 7.3.1). The semi-transparent arrows indicate the SFs for a flavour derived in one step being applied to rescale the distribution(s) of the same flavour of jets in the next step(s)

with the 2D maps of SFs for the CvsL–CvsB plane, give the full set of SFs required for the calibration of the DeepCSV c taggers.

DeepJet Since the CvsL = CvsB = −1 bin of DeepJet consists almost entirely of b jets, a lack of data for light-flavour and c jets in this bin makes it both impossible and unnecessary to perform a three-flavour iterative fit to find DeepJet SFs for udsg and c. Therefore, in this bin, c and udsg SFs are both set to 1 ± 1, and the b SF is evaluated as:

$$\mathrm{SF}_b^{-1} = \frac{N_{\mathrm{data}} - N_{\mathrm{sim,c}} - N_{\mathrm{sim,udsg}}}{N_{\mathrm{sim,b}}}, \tag{7.4}$$

where N represents the jet counts of the respective flavours in the bin containing jets with default discriminator value in the $t\bar{t}$ sample.

7.4 Interpolation

In this subsection, an interpolation scheme is discussed that combines the SF results obtained in two different binning schemes (as explained in Sect. 7.3) into one set of SFs with finer binning.

An interpolation of the SF values in the 2D plane is performed on a per-flavour basis to find more representative corrections on a finer grid. Such an interpolation is feasible because the SFs have a fairly smooth distribution, as evident to a first degree of approximation from the data-to-simulation ratios in Fig. 7.2. Moreover, combining the two sets of SFs, which are derived using two different binning schemes with such an interpolation approach, is expected to improve the SF values, given the fact that the two sets of SFs potentially contain complementary information about rapidly varying SF values at certain parts of the CvsL–CvsB plane. For example, the SF set derived with an adaptive binning scheme along the CvsB axis could potentially reflect sharp SF variations along the CvsB axis for a certain flavour of jets in parts of the plane enriched in jets of that flavour.

At first, a nodal point for the interpolation (a point where the functional value is assumed to be known exactly) is defined for each bin in the 2D plane for each flavour in each SF set. The coordinates of this nodal point are defined as the means of the coordinates of all the jets of the corresponding flavour that are contained in the corresponding bin. Thus, the nodal point (x, y) of a bin in an SF map for a given flavour containing N jets of that flavour, each with a (CvsL, CvsB) value of (x_i, y_i), is given by:

$$(x, y) = \left(\frac{\sum_{i=1}^{N} x_i}{N}, \frac{\sum_{i=1}^{N} y_i}{N} \right). \tag{7.5}$$

The value of the SF corresponding to this bin is assigned to this nodal point and is assumed to be known exactly at this point for the purpose of the interpolation.

For each flavour, the nodal points obtained from the two SF sets are merged together to form an irregular grid of points in the 2D plane; an SF value is assigned to each from the respective SF set. When the same nodal point is found in both the SF sets, only one is used to form the grid. A representative distribution of nodal points with their SF values in the 2D plane is shown in Fig. 7.5.

A bivariate interpolation is performed for each flavour using the SF values at the nodes of the corresponding irregular grid using the `interpolate` subpackage of SciPy [14]. The algorithm first triangulates the nodal points in the 2D plane using

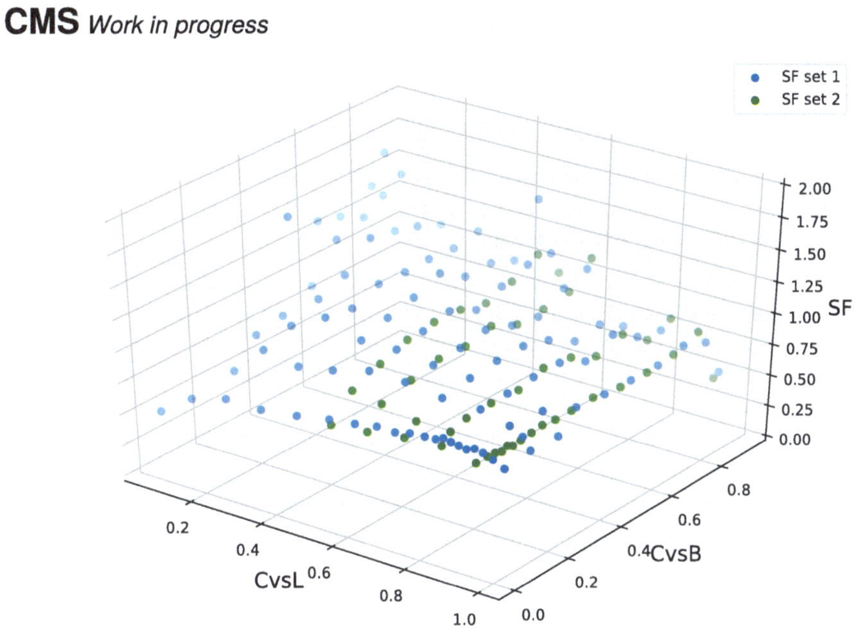

Fig. 7.5 An illustration showing several SF nodal points, taken from two sets of SFs with different binning schemes, distributed in the 2D CvsL–CvsB plane. The z-axis represents the SF value at each of these nodal points. "SF set 1" refers to the SFs derived with the CvsB axis as the fixed axis and CvsL as the variable axis, while "SF set 2" refers to the SFs derived with the opposite choice of axes. Points farther from the viewer appear lighter in colour

QHULL [15] and then constructs a piecewise cubic interpolating Bézier polynomial [16] on each triangle using a Clough-Tocher scheme [17]. Then the interpolant is constructed by choosing gradients that approximately minimize the curvature of the interpolating surface [18, 19].

However, such an interpolant is defined only inside the convex polygon (convex hull) containing all the nodal points. Furthermore, the interpolant is seen to have unphysical properties in some parts of the plane inside the convex hull but outside the concave hull of the nodal points. Therefore, a nearest-neighbour extrapolation (assigning the functional value of the nodal point nearest to a given point) is performed to assign values to all points in the 2D plane that lie outside the concave hull of the nodal points. The interpolant along with the extrapolant together define SF values at all points of the 2D plane (called interpolated SFs hereafter, for simplicity). Even though this allows for a continuous correction over all the plane, the interpolated SF values are computed at 50 discrete bins along each axis for convenience, which is a good approximation for a continuous correction.

It has been verified that interpolating using nodal points from two sets of SFs derived with two different binning schemes provides a final set of SFs that exhibits a better closure of the method than SFs derived with interpolation using nodal points from any one set alone, in case of both DeepCSV and DeepJet algorithms. Furthermore, it was also verified that the piecewise-cubic bivariate interpolant used here outperforms other methods of interpolation, such as piecewise linear bivariate interpolation and smoothing bivariate spline approximation.

7.5 Uncertainty Estimation

The statistical uncertainty in the SF for a given flavour in a given bin is evaluated from the statistical uncertainty in the number of jets in both data and simulation in that bin in the selection enriched in that flavour. In this regard, the per-bin uncertainty in simulation, as well as data, of the W+c OS-SS selection accounts for the individual statistical uncertainties in the OS and SS selections, and is hence propagated to the statistical uncertainties in the c jet SFs.

For systematic uncertainty estimation, all selections are performed with different sources of systematic uncertainties shifted by ± 1 standard deviation independently, and the whole iterative fit procedure is performed for every systematic variation. This gives modified SF maps for every source of uncertainty. The sources of systematic uncertainties considered are:

- Uncertainties in lepton identification uncertainties [13, 20, 21]
- Uncertainty in pileup modelling
- JES uncertainty [22, 23]
- JER uncertainty [22, 23]
- Uncertainties in factorisation and renormalisation scales in simulation
- Uncertainties in initial- and final-state radiation (FSR) in the parton shower
- Uncertainties in theoretical cross sections of the various physics processes used in the selections [2–7]
- Uncertainties in the b quark fragmentation modelled in PYTHIA
- Uncertainties in flavour compositions of V+jet processes
- Interpolation uncertainties
- Extrapolation uncertainties.

The uncertainty in flavour compositions is important as the method can potentially be sensitive to the flavour composition of the background jets. Additional uncertainties are assigned to the b and c jet yields in the DY+jet process and the c jet yield in the W+c process. These uncertainties are estimated to be 2.5 and 9%, respectively, for b and c jets produced in association with a Z boson, and 8% for c jets produced in association with a W boson; these results are based on the uncertainties in the experimental measurement of Z+b/c processes [11] and the experimental uncertainty in the measurement of the W+c cross section [8].

Although there is no straightforward method to estimate the bounds of the interpolation uncertainties at non-nodal points in the 2D plane in case of a piecewise-cubic bivariate interpolant, an approximate order-of-magnitude estimate is attempted. For each non-nodal point (probe point) in the concave hull of the SF nodal points for a given flavour, the interpolation uncertainty is quantified as the sum of 10 terms corresponding to the 10 control points of the cubic Bézier triangle that contains the point. Each of these terms is proportional to (a) the maximum absolute deviation of the directional derivative of the interpolant at each point between the control and probe points, from the average slope of the interpolant between the control and probe points, and (b) the linear distance between the control point and the probe point. This quantity is expected to be larger in parts of the plane where the interpolant has a high curvature, and at points farther away from the nodal points. In all cases, the quantity is computed approximately using numerical methods. This uncertainty is set to zero for points outside the concave hull of the nodal points.

On the other hand, the extrapolation uncertainty at a point outside the concave hull of the SF nodal points for a given flavour, is defined as the difference between the SF value obtained from a bicubic bivariate smoothing spline extrapolation [14] and the nominal SF value assigned (using nearest neighbour extrapolation). This uncertainty is set to zero for points inside the concave hull.

An estimation of the total uncertainty is made by adding statistical and all systematic uncertainties in quadrature, assuming no correlation. An estimation of the relative contributions of each source of uncertainty is presented in Sect. 7.6.

7.6 Scale Factors

The shape calibration SFs, obtained using the iterative fit method for DeepJet-based c taggers using the datasets corresponding to the 2017 data-taking conditions, are shown in Fig. 7.6. The set of SFs obtained using CvsB as the fixed axis and CvsL as the variable axis is shown in Fig. 7.6a. These set of SFs capture rapid variations along the CvsL axis. On the other hand, the set of SFs obtained using CvsL as the fixed axis and CvsB as the variable axis capture rapid variations along the CvsB axis, and have been shown in Fig. 7.6b. The effect is visible, for example, in the high-CvsL, low-CvsB regions (lower-right corner) of the b flavour SFs (middle plot of Figs. 7.6a and b). Using the interpolation scheme described in Sect. 7.4, these two sets of SFs are merged into one set, as shown in Fig. 7.6c.

The plots in Fig. 7.6, however, do not show the magnitude of uncertainties associated with the SFs at different parts of the plane. The effect of each source of systematic uncertainty in these SFs, when projected onto each of the 1D CvsL and CvsB distributions, is demonstrated by applying the nominal SFs separately to c, b, and light-flavour jets taken from a simulated hadronically decaying $t\bar{t}$ sample, and then measuring the relative changes in the jet yield distributions by applying each systematic variation of the SFs. The relative contributions of all systematic uncertainty sources thus projected for both the taggers and shown in Fig. 7.7, again for the DeepJet algorithm and the datasets corresponding to the 2017 data-taking conditions. Interpolation and extrapolation uncertainties have the highest contribution in the SFs in case of c jets. This is a result of the sparse binning of the nodal points in the c jet

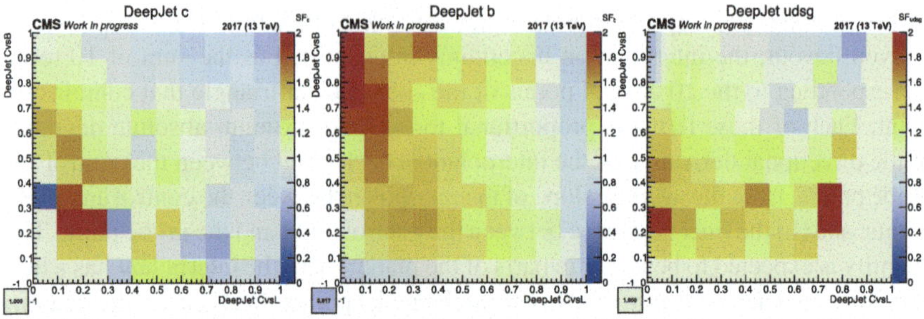

(a) With CvsB (y-axis) chosen as the fixed-width axis, and CvsL (x-axis) as the variable-width axis.

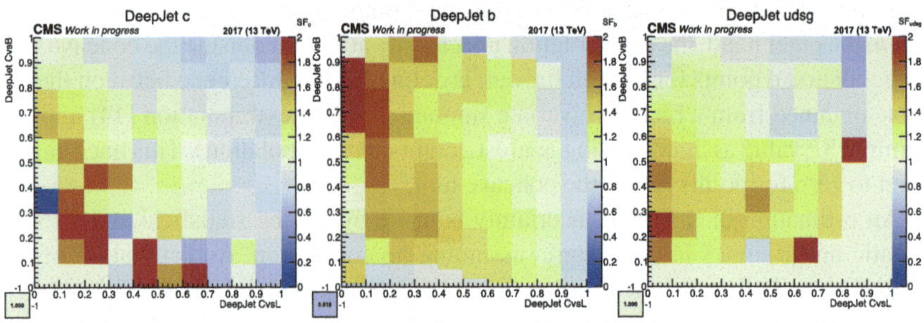

(b) With CvsL (x-axis) chosen as the fixed-width axis, and CvsB (y-axis) as the variable-width axis.

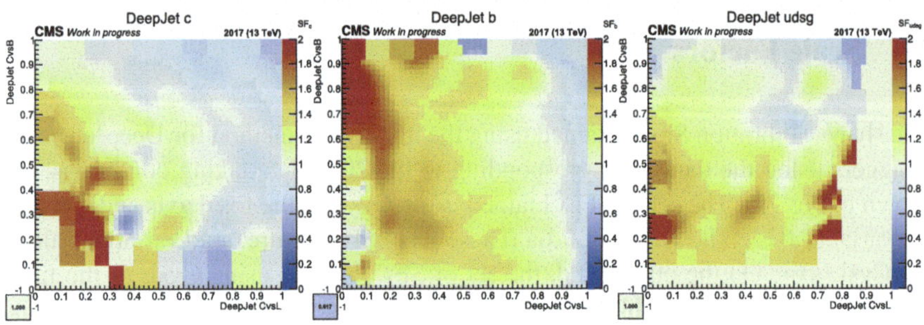

(c) Final SF results after interpolating using information from both binning schemes.

Fig. 7.6 The shape calibration SF values as a function of CvsL (x-axis) and CvsB (y-axis) for the DeepJet-based c taggers, for c jets (left), b jets (middle), and udsg jets (right), taken from the datasets corresponding to the 2017 data-taking conditions, are shown as heat maps. The lower-left box per plot represents the SF for jets with the default discriminator values. The z-axis represents the SF value

SFs all over the 2D plane, which, in turn, is a result of high statistical uncertainties in the W+c OS-SS selection that predominantly constrains the c jet SFs. Uncertainties in the factorisation scale alone dominate among the light-flavour SF uncertainties, especially in parts of the phase space having low yields for the corresponding flavour of jets. The uncertainties in the ISR and FSR in the PS have the highest contribution to the SF uncertainties for b jets, since these variations are relatively larger in the $t\bar{t}$ selections from which the SFs for b jets are predominantly constrained.

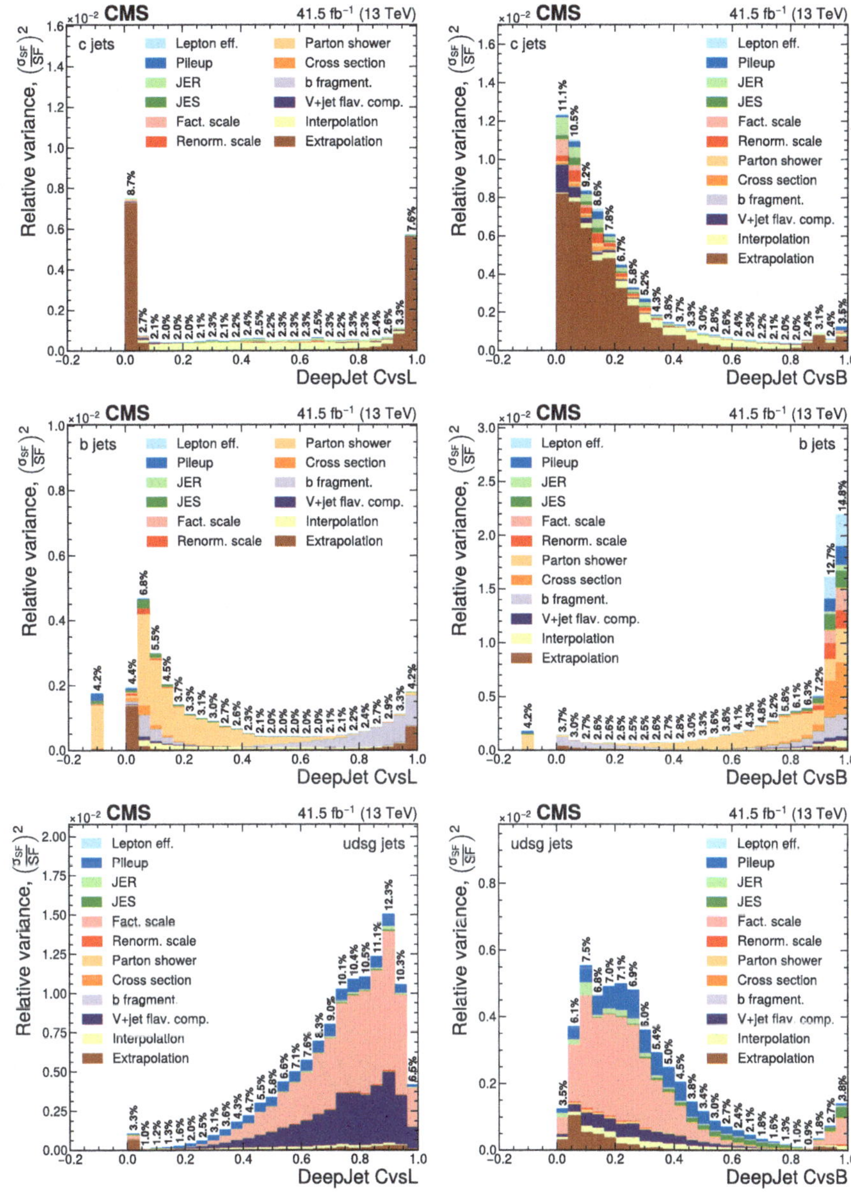

Fig. 7.7 Contribution of each source of the SF uncertainty, calculated as the square of the relative uncertainty in the jet yield and expressed as the maximum of the up and down variations, at various values of the DeepJet CvsL (left) and CvsB (right) discriminators for c (upper), b (middle), and light (lower) flavours, for the datasets corresponding to the 2017 data-taking conditions. The effective total relative uncertainty values ($\sqrt{\sum\left(\frac{\sigma_{SF}}{SF}\right)^2}$) per bin are also shown in bold text, for reference. The bin corresponding to a tagger value of -1 is plotted at -0.1. Statistical uncertainties are not shown

7.7 Adjusted Charm Tagging Performance in Data

The c tagging shape calibration SFs are used to adjust the tagger discriminator distributions of simulated jets of each flavour to obtain an estimate of the corresponding distributions of jets in data. This is performed using an independent sample of jets taken from simulated hadronically decaying $t\bar{t}$ events and the adjusted distributions are then used to plot the ROC curves to estimate c tagging performance in data. Furthermore, the statistical and systematic uncertainties in the SFs are suitably propagated to the ROC curves. The adjusted ROC curves for CvsL and CvsB discrimination along with their variations because of statistical and systematic uncertainties are shown in Fig. 7.8. In a similar fashion, the adjusted c tagging efficiencies, as functions of the adjusted b and light-flavour jet misidentification rates are shown in Fig. 7.9. A comparison with DeepCSV c taggers has also been shown in the same figures. DeepJet outperforms DeepCSV in both CvsL and CvsB performances in the adjusted ROCs in data. This motivates the use of DeepJet as the preferred charm tagging algorithm in the analyses presented in the following sections.

Individual contributions from each source of uncertainty are also quantified by the change in the area under the ROC curve from that of the central ROC. The relative contributions, including that of statistical uncertainties, evaluated using the ROC curve variations as described here are presented graphically in Fig. 7.10. The approach of presenting uncertainties as a function of discriminator values in Fig. 7.7 is essentially different from this interpretation, because the former disregards the relative abundance of jets at different discriminator values, whereas the latter includes

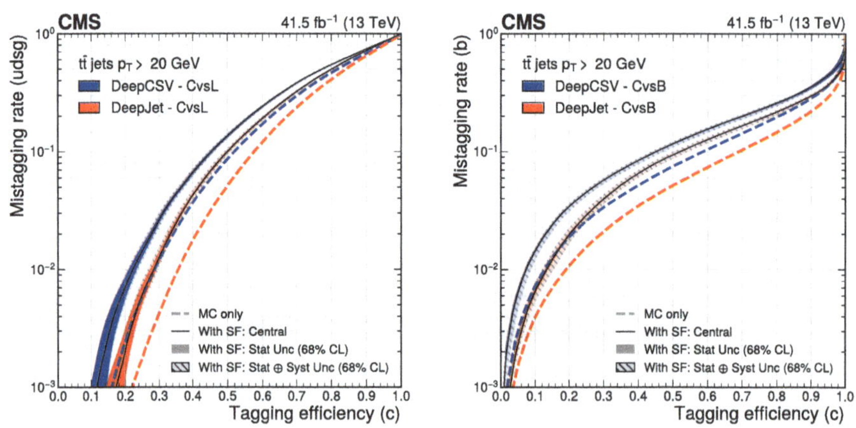

Fig. 7.8 The ROC curves showing the individual performance of the CvsL (left) and CvsB (right) discriminators for the DeepCSV (blue) and DeepJet (red) algorithms for simulated jets (the dashed lines) and the estimation of the same for jets in data (the solid lines), for the datasets corresponding to the 2017 data-taking conditions. The solid uncertainty bands around the solid lines represent statistical uncertainties only, and the hatched semi-transparent bands represent statistical and systematic uncertainties added in quadrature

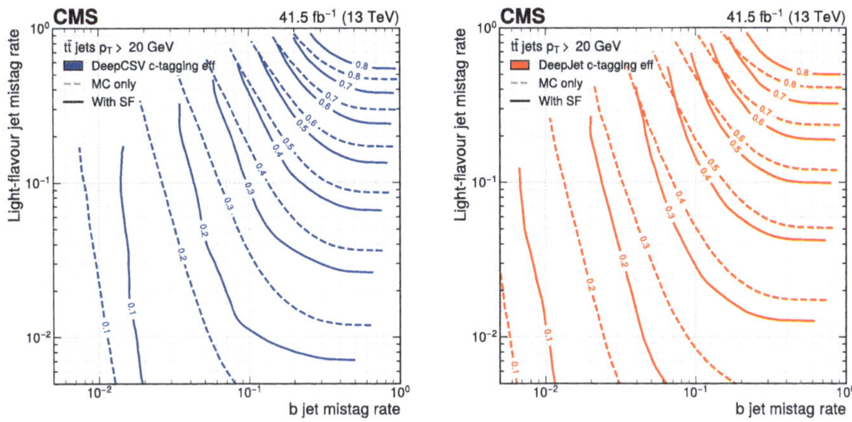

Fig. 7.9 The ROC contours showing c tagging efficiencies as functions of b and light-flavour jet misidentification rates, for the DeepCSV (left) and DeepJet (right) algorithms for simulated jets (the dashed lines) and the estimation of the same for jets in data (the solid lines), for the datasets corresponding to the 2017 data-taking conditions. Each line represents points in the plane that correspond to a fixed value of the c tagging efficiency, which is shown as a number at the centre of each line

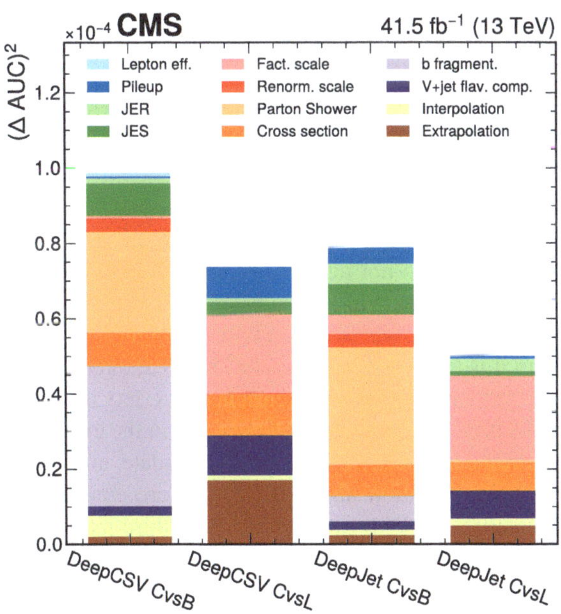

Fig. 7.10 Relative contributions of each source of uncertainty to the total uncertainty (statistical + systematic) for both CvsL and CvsB discrimination and for both DeepCSV and DeepJet taggers, quantified by the square of the change in area under ROC curves

the absolute change in the jet yields in different parts of the phase space and could therefore be more representative of the corresponding effect in a physics analysis. In this approach, uncertainties in the factorisation scale are the dominant contributions to the overall uncertainties in CvsL discrimination, whereas uncertainties in b fragmentation and in the ISR and FSR in the PS are among the highest contributors to the uncertainties in the CvsB discriminator, owing to their large contributions to the b jet SFs.

7.8 Validation

7.8.1 Closure Test

A closure test is performed by applying the derived shape calibration SFs to the CvsL and CvsB distributions of the same jet selections for which the SFs were derived. The effect of applying the DeepJet SFs on the DeepJet c tagger distributions in the selections with the datasets corresponding to the 2017 data-taking conditions is shown in Fig. 7.11. A good agreement between the c tagging discriminator distributions of simulated jets and those of jets in data establishes the closure of this method.

7.8.2 Validation on Jets Not Biased with Muons

Two additional tests are performed to demonstrate applicability of the SFs, which have been derived using b and c jets containing a soft muon inside them, when applied to jet samples that do not necessarily contain a soft muon. To construct a sample of jets enriched in b jets without using soft muons to identify them, the same $\mathrm{t\bar{t}}$ dileptonic selection is used, where the jet with the highest p_{T} that was not tagged with a soft muon in it is treated as the "second" b jet candidate.

Similarly, a second sample relatively enriched in c jets is constructed from the $\mathrm{t\bar{t}}$ semileptonic selection. All possible triplets of jets in the events are considered to reconstruct the hadronically decaying W boson candidate, and hence, the top quark candidate. For each combination, the first two jets are considered as the c and s candidates to reconstruct the W boson, whereas the third jet is considered as the b jet candidate. All combinations where the jet tagged with a soft muon is selected as the c jet or s jet candidate are rejected. For all other combinations, a mass-based χ^2 is defined as

$$\chi^2 = \left(\frac{m_{\mathrm{W}} - 80.3}{20}\right)^2 + \left(\frac{m_{\mathrm{t}} - 172.5}{30}\right)^2, \tag{7.6}$$

with m_{W} and m_{t} as the invariant masses of the reconstructed W boson and top quark candidates, respectively, in GeV. The combination of jets that yields the lowest value

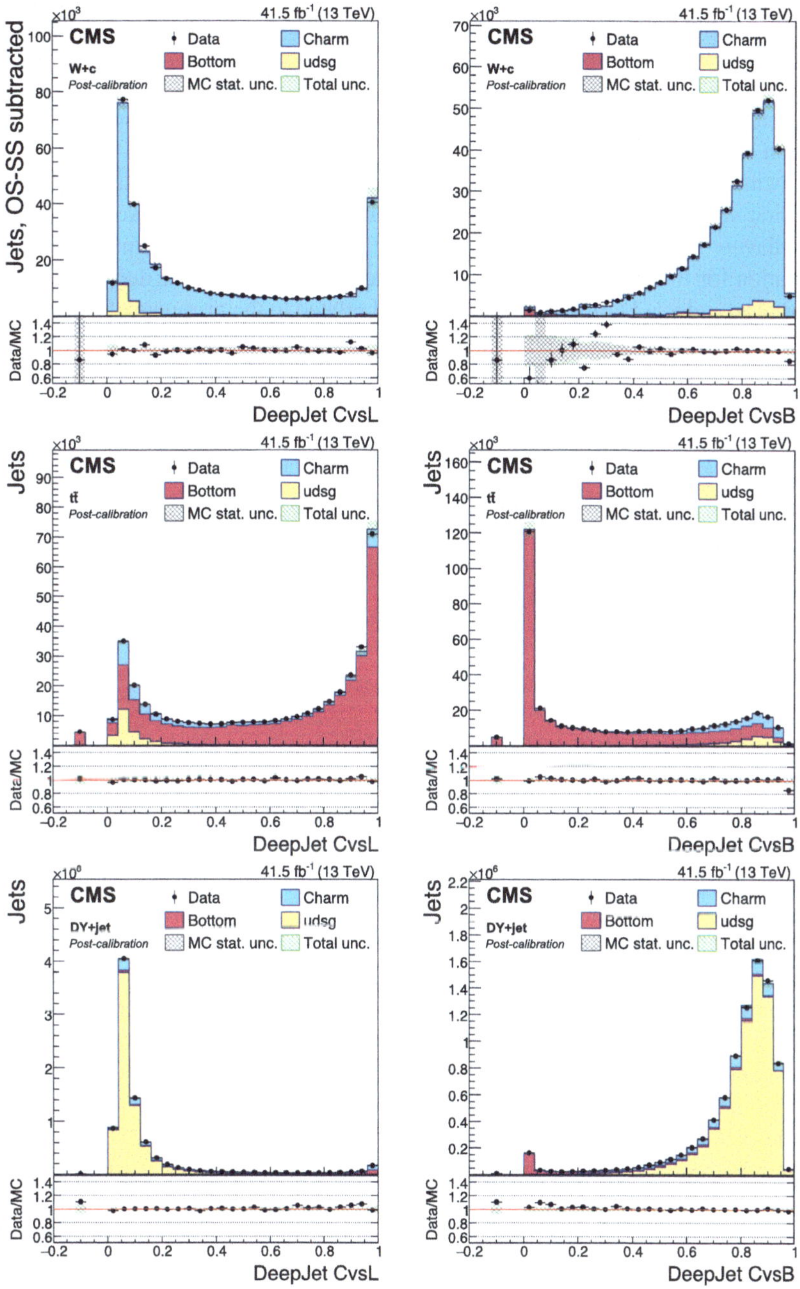

Fig. 7.11 Post-calibration DeepJet CvsL (left) and CvsB (right) distributions of jet samples selected from W+c (upper), $t\bar{t}$ semi- and dileptonic (middle), and DY+jet (lower) events from the datasets corresponding to the 2017 data-taking conditions, after application of DeepJet c tagger shape calibration SFs. The bin corresponding to a tagger value of −1 is plotted at −0.1. Vertical error bars in data represent statistical uncertainties in data

of χ^2 for a given event is considered the best combination, and the highest-p_T jet assigned to the hadronically decaying W boson candidate is considered the c jet candidate.

Each jet in both these samples may or may not contain a soft muon, and the proportion of jets with and without a soft muon should be distributed as it would be in any inclusive selection of jets with similar event selections and flavour composition. The first sample is expected to contain mainly b jets and light-flavour jets. Since light-flavour jet SFs are already known to be free from any bias from soft muons, a validation for this sample would demonstrate applicability of the derived SFs on b jets that are selected without any requirement related to the presence of a soft muon. The second sample, on the other hand, is expected to contain c jets with significant contamination from b and light-flavour jets. A validation using this sample, in combination with the conclusion from the aforementioned validation on b jets, would demonstrate the applicability of the SFs for all c jets as well.

The pre- and post-calibration distributions of the DeepJet c tagging discriminator values of the two samples of jets thus selected from the datasets corresponding to the 2017 data-taking period are shown in Figs. 7.12 (dileptonic $t\bar{t}$) and 7.13 (semileptonic $t\bar{t}$). A much flatter data-to-simulation ratio is achieved after the corrections are applied, compatible with unity within a few percent for most of the discriminator values.

7.8.3 *Fit Validation with Pseudodata*

A test is performed to validate the accuracy of the SF values from the iterative convergence. A set of distributions of "pseudodata" is created by applying artificial, handcrafted SF maps to simulated jet distributions. To mimic the statistical independence of data from simulation, the central value of every bin of the pseudodata is set by sampling a random number from a normal distribution, whose mean is equal to the central value of the corresponding bin in the rescaled simulation and standard deviation is equal to the statistical uncertainty in the same bin. The relative uncertainty of every bin in the pseudodata, in turn, is set to the relative uncertainty in the corresponding bin of data. The iterative fit algorithm is now allowed to run on the same selections replacing data with pseudodata.

The artificial SFs are generated as an arbitrary, continuous 2D function of DeepCSV CvsL and CvsB values, individually for each jet flavour. A single arbitrary number is assigned as the SF for jets with the default DeepCSV discriminator value for each flavour. Subsequently, the SF map of a given flavour is normalised in such a way that the total jet yield of an independent selection of simulated jets of that flavour taken from a hadronically decaying $t\bar{t}$ sample remains unchanged upon application of the SF map under consideration.

Two such maps of artificial SFs are designed, the first (second) of which consists of "mild" ("strong") SFs, that is, SFs that are close to (far from) 1, with values in the range 0.5–1.4 (0.1–3.0). For each map, a total of 50 pseudodata models are created

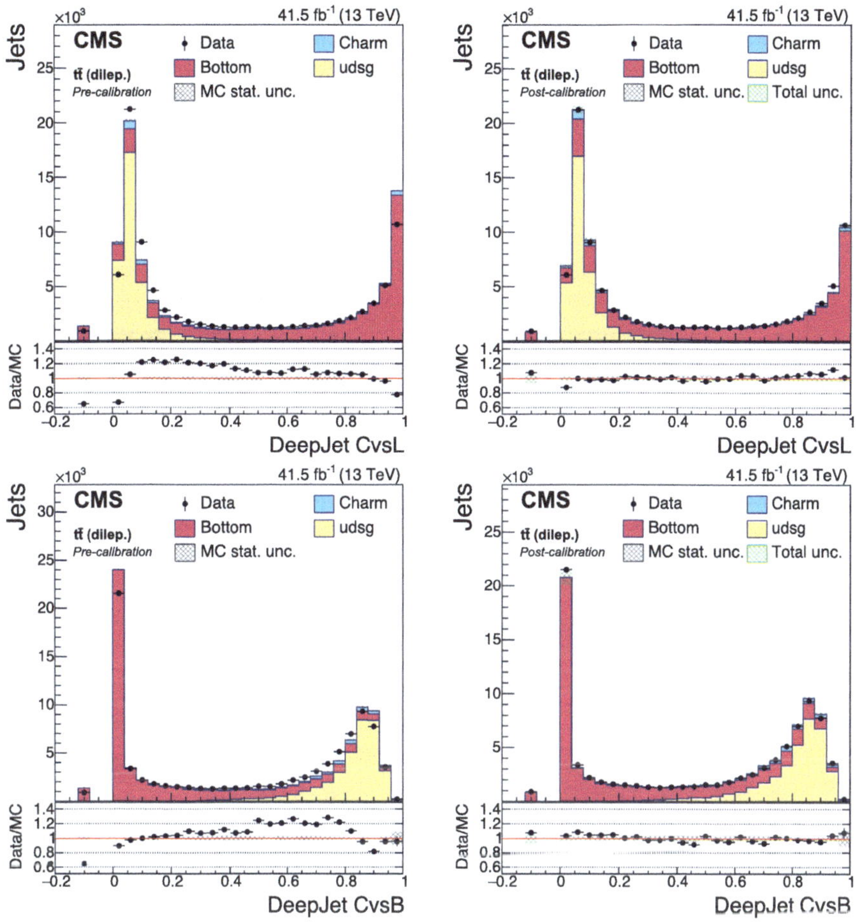

Fig. 7.12 DeepJet CvsL (upper) and CvsB (lower) discriminators of dileptonic $t\bar{t}$ jets not biased with soft muons, before (left) and after (right) application of SFs. The bin corresponding to a tagger value of -1 is plotted at -0.1. Vertical error bars in data represent statistical uncertainties in data

injecting the same set of artificial SFs but using a different random number seed in each model. For each map, the difference between the injected SFs and the extracted SFs ("SF pull") is plotted in Fig. 7.14. The differences are measured in units of the statistical uncertainty of the extracted SFs across all bins, weighted by the relative jet abundance in each bin in the CvsL–CvsB plane. The figure shows the cumulative results from all 50 models for c, b, and light favours separately. A mean of around 0 of the SF pulls illustrates a correct convergence of the method. A standard deviation of around 1 illustrates an optimal estimation of the SF statistical uncertainties in most cases. A standard deviation smaller than 1 for b jets in case of the "mild" map illustrates a conservative estimation of SF statistical uncertainties in case of "mild" deviations from data in simulation.

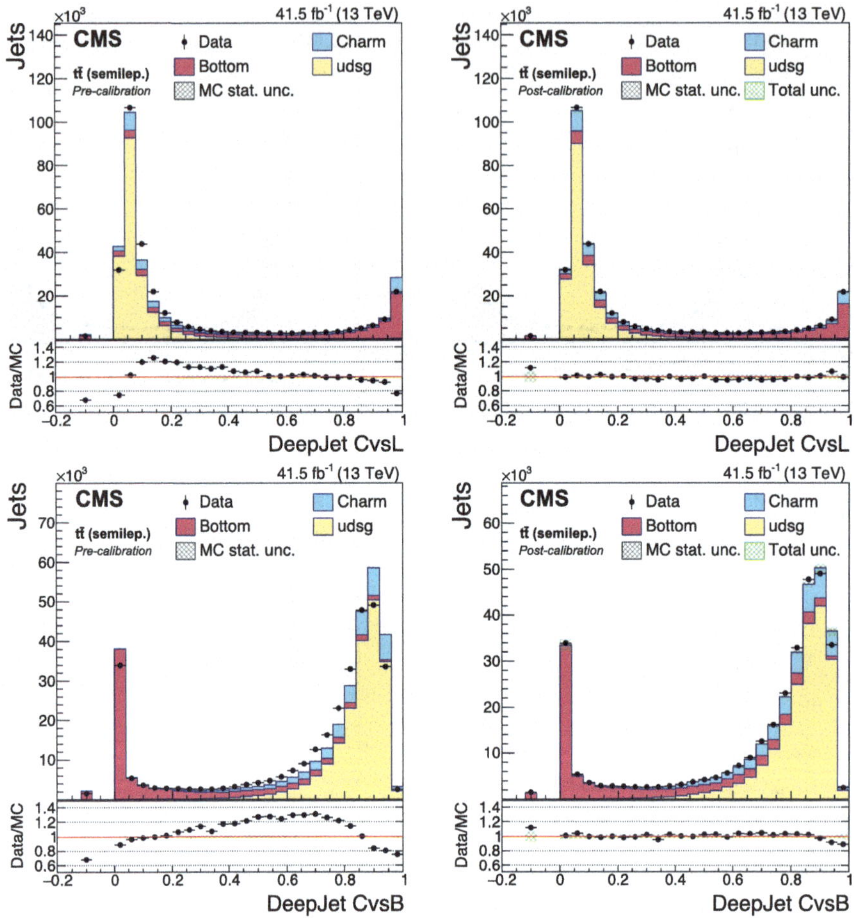

Fig. 7.13 DeepJet CvsL (upper) and CvsB (lower) discriminators of semileptonic $t\bar{t}$ jets not biased with soft muons, before (left) and after (right) application of SFs. The bin corresponding to a tagger value of -1 is plotted at -0.1. Vertical error bars in data represent statistical uncertainties in data

Whereas these studies show that the statistical uncertainties might be somewhat conservative in regions of the phase space where SFs for b jets are close to 1, this might not be the case for other regions where SFs show larger deviations. Since the SFs that map simulated b jets to b jet candidates in collision data (Fig. 7.6) are measured to have high central values close to 2 at some parts of the phase space, no a posteriori reduction of the uncertainties has been applied in the analyses in the following section.

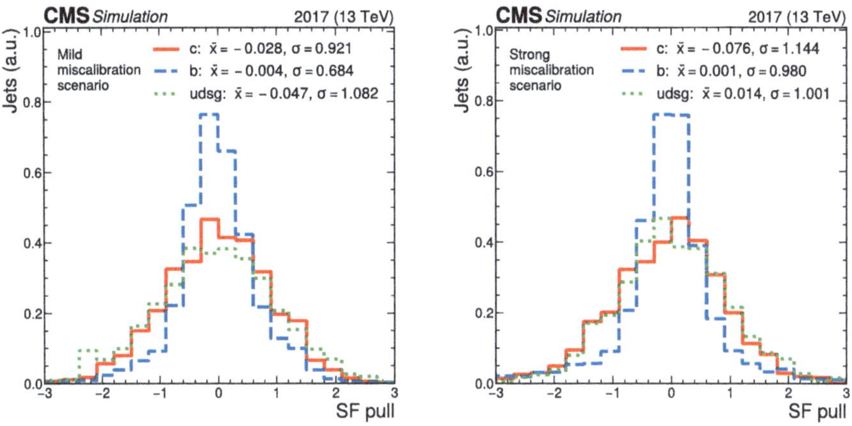

Fig. 7.14 Distribution of the SF pulls, quantified as the differences between the injected SFs and the SFs retrieved by the fits in units of the statistical uncertainties in the latter ($\frac{SF_{extracted}-SF_{injected}}{\sigma_{extracted}}$), across all bins in the CvsL–CvsB plane, for the SF map with "mild" (left) and "strong" (right) SFs

7.9 Summary of the Calibration

This section presents a novel method to calibrate the full differential shape of the discriminator distributions used for charm jet identification at CMS, closely following Ref. [1]. The method uses three different sets of event selection criteria, targeting topologies enriched in W+c, top quark pairs, and DY+jet events. These topologies are highly enriched in c, b and light-flavour jets, respectively, resulting in purities of a given jet flavour that range between 81 and 93%. By employing an iterative fitting approach in each of these three regions, scale factors (SFs) are derived to match the simulated discriminator distributions to those observed in data. Since the c tagging algorithm is composed of two discriminators, one to discriminate c from b jets (CvsB) and another to discriminate c from light-flavour and gluon jets (CvsL), the SFs are derived as functions of CvsL and CvsB discriminator values. An adaptive binning is used to optimise the granularity of the provided calibration with respect to the statistical uncertainty in each bin. Finally, an interpolation is performed to obtain more representative corrections over the entire two-dimensional plane.

Validation and closure tests confirm the robustness of the method. Although this section reports calibration results of the DeepJet c taggers with only data and simulated samples corresponding to the 2017 data-taking conditions, similar calibrations are obtained with those corresponding to the 2016 and 2018 data-taking conditions separately and are used for the analysis of data collected in the respective years. The calibration of the full differential discriminator shape allows the use of the c tagging discriminators as inputs to multivariate techniques (based on machine learning) or by fitting the discriminator shapes to data to extract observables that are sensitive to the jet flavour. The shape calibration extends the use of c tagging algorithms beyond the application of discrete working points, and facilitates more advanced uses for c jet identification in physics analyses, as has been demonstrated in the analyses presented in the following sections.

References

1. CMS Collaboration (2022) A new calibration method for charm jet identification validated with proton-proton collision events at $\sqrt{s}$ =13 TeV. JINST 17:P03014. https://doi.org/10.1088/1748-0221/17/03/P03014. arXiv:2111.03027
2. Li Y, Petriello F (2012) Combining QCD and electroweak corrections to dilepton production in FEWZ. Phys Rev D 86:094034. https://doi.org/10.1103/PhysRevD.86.094034. arXiv:1208.5967
3. Frixione S et al (2008) Single-top hadroproduction in association with a W boson. JHEP 07:029. https://doi.org/10.1088/1126-6708/2008/07/029. arXiv:0805.3067
4. Re E (2011) Single-top Wt-channel production matched with parton showers using the POWHEG method. Eur Phys J C 71:1547. https://doi.org/10.1140/epjc/s10052-011-1547-z. arXiv:1009.2450
5. Kidonakis N (2014) NNLL threshold resummation for top-pair and single-top production. Phys Part Nucl 45:714. https://doi.org/10.1134/S1063779614040091. arXiv:1210.7813
6. Campbell JM, Ellis RK (2010) MCFM for the Tevatron and the LHC. Nucl Phys Proc Suppl 205–206:10. https://doi.org/10.1016/j.nuclphysbps.2010.08.011. arXiv:1007.3492
7. Gehrmann T et al (2014) W^+W^- production at hadron colliders in next to next to leading order QCD. Phys Rev Lett 113:212001. https://doi.org/10.1103/PhysRevLett.113.212001. arXiv:1408.5243
8. CMS Collaboration (2019) Measurement of associated production of a W boson and a charm quark in proton-proton collisions at $\sqrt{s} = 13$ TeV. Eur Phys J C 79:269. https://doi.org/10.1140/epjc/s10052-019-6752-1. arXiv:1811.10021
9. Collaboration CMS (2021) Measurement of differential $t\bar{t}$ production cross sections in the full kinematic range using lepton+jets events from proton-proton collisions at $\sqrt{s} = 13$ TeV. Phys Rev D 104:092013. https://doi.org/10.1103/PhysRevD.104.092013. arXiv:2108.02803
10. CMS Collaboration (2019) Search for new physics in top quark production in dilepton final states in proton-proton collisions at $\sqrt{s} = 13$ TeV. Eur Phys J C 79:886. https://doi.org/10.1140/epjc/s10052-019-7387-y. arXiv:1903.11144
11. CMS Collaboration (2020) Measurement of the associated production of a Z boson with charm or bottom quark jets in proton-proton collisions at $\sqrt{s} = 13$ TeV. Phys Rev D 102:032007. https://doi.org/10.1103/physrevd.102.032007. arXiv:2001.06899
12. CMS Collaboration (2016) Identification of c-quark jets at the CMS experiment. CMS Physics Analysis Summary. CMS-PAS-BTV-16-001. http://cds.cern.ch/record/2205149
13. CMS Collaboration (2021) Electron and photon reconstruction and identification with the CMS experiment at the CERN LHC. JINST 16:P05014. https://doi.org/10.1088/1748-0221/16/05/p05014. arXiv:2012.06888
14. Virtanen P et al (2020) SciPy 1.0: fundamental algorithms for scientific computing in Python. Nat Methods 17:261. https://doi.org/10.1038/s41592-019-0686-2
15. Barber CB, Dobkin DP, Huhdanpaa H (1996) The quickhull algorithm for convex hulls. ACM Trans Math Softw 22:469. https://doi.org/10.1145/235815.235821
16. Bézier P (1970) Numerical control: mathematics and applications. Wiley, London
17. Clough R, Tocher J (1965) Finite element stiffness matricess for analysis of plate bending. In: Proceedings of the first conference on matrix methods in structural mechanics, p 515
18. Nielson GM (1983) A method for interpolating scattered data based upon a minimum norm network. Math Comput 40:253. https://doi.org/10.1090/S0025-5718-1983-0679444-7

19. Renka RJ, Cline AK (1984) A triangle-based C^1 interpolation method. Rocky Mt J Math 14:223. https://doi.org/10.1216/RMJ-1984-14-1-223
20. CMS Collaboration (2018) Performance of the CMS muon detector and muon reconstruction with proton-proton collisions at $\sqrt{s} = 13$ TeV. JINST 13:P06015. https://doi.org/10.1088/1748-0221/13/06/P06015. arXiv:1804.04528
21. CMS Collaboration (2020) Performance of the reconstruction and identification of high-momentum muons in proton-proton collisions at $\sqrt{s} = 13$ TeV. JINST 15:P02027. https://doi.org/10.1088/1748-0221/15/02/P02027. arXiv:1912.03516
22. CMS Collaboration (2018) Jet energy scale and resolution performance with 13 TeV data collected by CMS in 2016. CMS Detector Performance Note. CMS-DP-2018-028. https://cds.cern.ch/record/2622157
23. CMS Collaboration (2019) Performance of missing transverse momentum reconstruction in proton-proton collisions at $\sqrt{s} = 13$ TeV using the CMS detector. JINST 14:P07004. https://doi.org/10.1088/1748-0221/14/07/P07004. arXiv:1903.06078

Chapter 8
Analysis Strategy

This thesis discusses three physics analyses performed using pp collision data collected by the CMS experiment at a centre of mass energy ($\sqrt{s}$) of 13 TeV, between 2016 and 2018. The analyses are:

1. **VHcc search**: a search for the Higgs boson decaying to a charm quark-antiquark pair in events where the Higgs boson is produced in association with a vector (W or Z) boson
2. **VZcc search**: a search for the Z boson decaying to a charm quark-antiquark pair in diboson (WZ and ZZ) events
3. **VWcq search**: a search for the W boson decaying to a pair of quarks of which one is a charm quark, in diboson (WW and WZ) events. Such processes are dominated by $W \rightarrow cs$ decays.

The signal processes for each of these searches are discussed in Sects. 2.3 and 2.4. All three searches make use of charm (c) tagging and the novel calibration method, discussed in Sects. 6 and 7, respectively. This section discusses the overall analysis strategy used in these three searches.

8.1 Search Strategy

The general strategy in each of the three analyses is to reconstruct a leptonically-decaying vector boson (W or Z boson) recoiling against a Higgs, Z, or W boson, that undergoes a charmed decay. The hadronically-decaying boson is reconstructed based on its decay into a pair of AK4 jets that are likely to have been initiated by c quark(s). Depending on the type and decay mode of the leptonically-decaying vector boson, events are classified into three categories:

S. Mondal, *Charming Decays of the Higgs, Z, and W Bosons*, Springer Theses,
https://doi.org/10.1007/978-3-031-84677-9_8

- **0 lepton channel (0L)**, to reconstruct the Z boson decaying into a pair of neutrinos (Z $\rightarrow \nu\nu$), i.e. no charged leptons in the final state
- **1 lepton channel (1L)**, to reconstruct the W boson decaying into a charged lepton (electron or muon) and a neutrino (W $\rightarrow \ell\nu$), i.e. one charged lepton in the final state
- **2 lepton channel (2L)**, to reconstruct Z boson decaying into a pair of oppositely charged electrons or muons (Z $\rightarrow e^+e^-/\mu^+\mu^-$), i.e. two same-flavour charged leptons in the final state.

Each of the channels is defined by the presence of the requisite number of reconstructed charged leptons. The 2L and 1L channels are further divided in two, depending on the flavour of the lepton (electron and muon channels). In the VHcc and VZcc searches, the 2L channel is further split into high-p_T (2LH) and low-p_T (2LL) regions, depending on the p_T of the reconstructed leptonically-decaying vector boson candidate. This is done to model the major backgrounds more accurately in each of these two channels individually, independent of each other.

The charmed decay of the Higgs/Z/W boson is reconstructed using two well-defined AK4 jets. Charm tagging criteria are used to suppress major backgrounds and enrich the signal selection (Signal Region or SR) with more events containing charmed decays. A BDT is trained to discriminate VHcc signal from background processes in the VHcc search, and VZcc processes from backgrounds in the VZcc search. The BDT uses the reconstructed mass and kinematic properties of the Higgs/Z boson, as well as the charm tagging scores of the two jets, as inputs. While the Higgs (Z) boson mass reconstructed from the two jets provides a strong discrimination against non-resonant backgrounds and dijet resonances at other masses, the c tagging scores provide a discrimination against H $\rightarrow$ b$\bar{\text{b}}$ (Z $\rightarrow$ b$\bar{\text{b}}$ and Z$\rightarrow q\bar{q}$, where q is a light quark) processes. This information along with other kinematics of the events helps the BDT discriminate signal from the backgrounds.

The VWcq search, instead, uses the invariant mass (m_{jj}) of the reconstructed charmed W boson as the discriminating variable. A BDT based on the c tagging scores of the jets is used a *priori* to discard events that are likely to contain two c jets (VZcc backgrounds in this search, for example), as discussed in Sect. 8.2.

Depending on the channel, Z+jets, W+jets, and t$\bar{\text{t}}$ processes form major, irreducible backgrounds. Such major background processes are estimated in a data-driven way using dedicated Control Regions (CRs) enriched with such processes. Furthermore, each of the simulated V+jets processes is separated into b, c, and light components, depending on the flavour of the selected jets, and the contribution of each component of simulation is estimated individually from data. This ensures that any mismodelling in the flavour composition of simulated V+jets events does not affect the results. Further details are provided in Sect. 9.6.

Minor backgrounds like single-top (ST) are estimated from the simulation. All diboson processes, along with the Higgs boson decaying to bottom quarks, form a minor background in the VHcc search. These are also estimated from simulation. In the VZcc and VWcq search, the diboson processes that do not include charmed decays constitute minor backgrounds and are estimated from simulation.

Finally, a binned maximum likelihood fit is performed using the distributions of the respective BDTs (m_{jj}) in the SR for the VHcc and VZcc (VWcq) search, simultaneously with charm tagging scores of jets in the CRs to extract the respective signal. While the CL_S metric is used to set an upper limit to the signal strength in the VHcc search, the significances of the signals are calculated in the VZcc and VWcq search. These statistical methods have been discussed in Sect. 4.7.2.

8.2 Correlated Charm-Strange Tagging for W Decays

As discussed in Sect. 6, c jets are identified by simultaneously discriminating them against light-flavour and b jets. Light-flavour jets, in turn, are defined as a combination of u, d, s, and gluon-initiated jets. The current multiclassification-based taggers in CMS classify strange-quark initiated jets under the udsg (DeepCSV) or uds (DeepJet) category. While it is possible to split the output classes to introduce an additional "s" class for dedicated strange tagging, such a discriminator has been seen to have a poor separation against udg jets. Moreover, a pure selection of strange jets in data needed for strange tagger calibration is also difficult to obtain. Hence, this thesis utilises the charm tagging scores to indirectly tag s jets.

Although the udsg output node of DeepJet is meant to include strange jets as well, true strange jets are expected to have a different CvsL (and CvsB) distribution compared to true udg jets as well as true charm jets, owing to their expected properties being intermediate to that of lighter ud and heavier c jets. Thus, the jet with the second-highest probability of being a c jet ("subleading jet") in VWcq events is a good candidate for an s jet. One could, in principle, use the CvsL score of this subleading jet as a parameter to discriminate VZcc against VWcq. However, since the true s jet (or other light-flavour jets in the event) may also have a CvsL score higher than that of the true c jet in VWcq processes, the CvsL and CvsB scores of both the leading and subleading jets have different distributions for VZcc and VWcq processes, as seen in simulated events shown in Fig. 8.1. Therefore, a BDT has been trained using the four scores (CvsL and CvsB scores of the leading and subleading jets) as inputs to discriminate VWcq events against VZcc events in the VWcq analysis.

Although this approach does not involve a direct s tagging algorithm (or a calibration of the same, thereafter), it is able to exploit the correlations between the calibrated c tagging scores of the c and s jet candidates of VWcq events to distinguish them from those of VZcc events. This indirect approach of "cs" tagging using c tagging scores of jet pairs has been referred to as correlated charm-strange (cs) tagging in this thesis and has been implemented in the VWcq analysis.

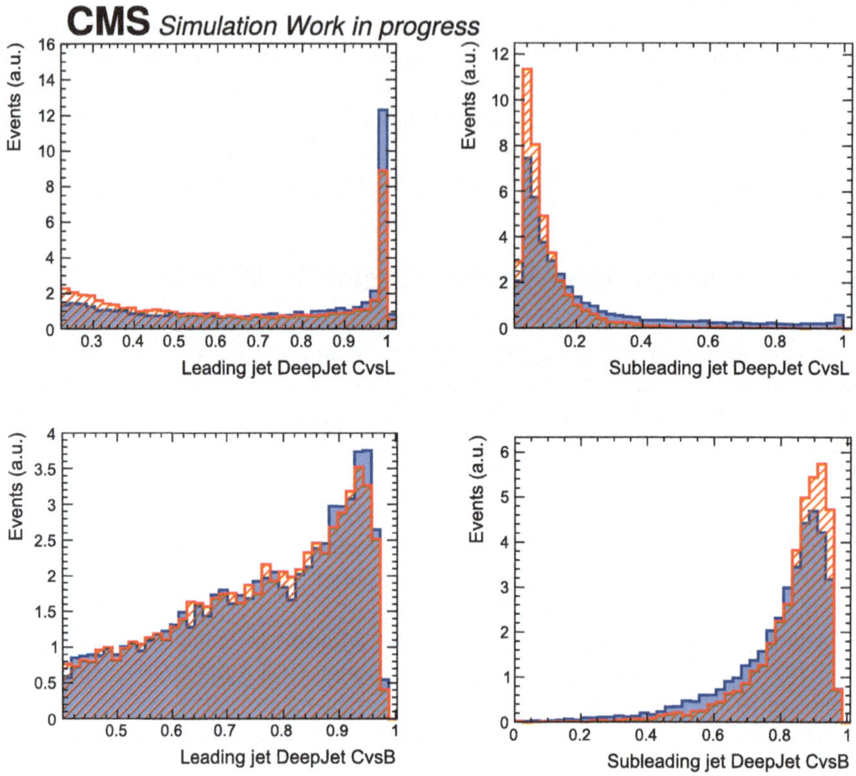

Fig. 8.1 DeepJet CvsL (upper) and CvsB (lower) distributions of the jet with the highest CvsL score in the event (leading jet, left) and the jet with the second-highest CvsL score in the event (subleading jet, right) for the VZcc process (blue) and VWcq process (red, hatched) in the 0L channel of the samples simulated using 2017 data-taking conditions. All four distributions provide some discrimination between the two processes. A BDT combining all this information further allows exploiting correlations between these distributions

8.3 The Merged-Jet Approach

The focus of this thesis is the reconstruction of the charmed Higgs/Z/W boson decay using two distinct AK4 jets. This approach is referred to as the resolved-jet approach. Another complementary approach to search for dicharm decays is to use a single, large-radius AK jet that contains both the charm quark decays within the jet cone. In particular, AK jets with a distance parameter of $\Delta R = 1.5$, known as AK15 jets, have been used in Refs. [1, 2] to reconstruct the Higgs and Z boson decay. Such an approach is referred to as the merged-jet approach. While the merged-jet approach is not the focus of this thesis, a brief summary of the methods and results from the approach is presented in this and following sections for completeness.

In the merged-jet approach, the state-of-the-art ParticleNet (PN) jet flavour tagging algorithm [3, 4] has been used to tag large-radius dicharm jets and distinguish them from light-flavour jets. The new PN tagger considers a jet as an unordered set of its constituent particles, effectively a "particle cloud". A customized neural network architecture using Dynamic Graph Convolutional Neural Network (DGCNN) is trained to distinguish dicharm AK15 jets in simulated H $\rightarrow$ cc events from light-flavour AK15 jets in simulated QCD multijet events. Such a tagger is able to exploit kinematic correlations between the constituents of two c quarks and thereby provide a stronger discrimination against backgrounds, compared to separately tagging two AK4 jets, each of which contains information about one c quark. The PN tagger cc tagging efficiency is calibrated at fixed working points (WPs) using the so-called sfBDT method [5], that uses a BDT to select a subset of $g \rightarrow c\bar{c}$ AK15 jets that mimic H $\rightarrow c\bar{c}$ AK15 jets in terms of the PN cc tagger score distribution, as a proxy for H $\rightarrow c\bar{c}$ jets. Comparing the distributions of such proxy jets in data and simulation, the efficiencies of cc tagging at different WPs in data and simulation are evaluated. Their ratios give a cc tagger SF that can be used to rescale simulated cc events to match those in data.

The merged-jet approach also considers leptonic decays of vector bosons in three channels (0L, 1L, and 2L). The Higgs/Z boson decay is reconstructed using one AK15 jet. Both the SRs and CRs are defined by requiring a stringent selection criteria on the PN cc-tagger score. Thereafter, a kinematic BDT is trained to distinguish signal events from background ones. Based on the BDT outputs of events, more signal-like events are assigned to the SR, while others are assigned to the CR. The BDT uses only kinematic variables of the reconstructed jets and vector bosons, without using the flavour tagging information or the invariant mass of the AK15 jet. Therefore, the SR and CR have comparable flavour compositions in each channel. Furthermore, an additional $t\bar{t}$ CR is defined by requiring additional AK4 jets in the event. The cc signal events are rescaled based on the SFs obtained from the sfBDT method.

The mass of jets is usually quantified using the so-called "Soft Drop" mass [6], m_{SD}, which in this case would quantify the mass of the Higgs/Z boson candidate. However, an additional AK15 jet mass regression [7] is performed, in order to assign a mass, m_{reg}, that is closer to the true mass of the particle that initiated the jet. This regression is performed using a DGCNN with the same architecture and inputs as the PN algorithm, and is trained in a "regression" mode with the generated particle mass as the target, using simulated events containing pseudo-Higgs bosons with a flat spectrum of masses between 15–250 GeV. Finally, the AK15 jet m_{reg} is used as the fit variable in a binned maximum likelihood fit to extract the VHcc or VZcc signal. As the PN tagger score, the kinematic BDT score, and the AK15 m_{reg} values contain more-or-less orthogonal information, the m_{reg} value provides additional discrimination against backgrounds even after the selection criteria on the tagger score and the kinematic BDT, and is therefore a suitable choice as the fit variable. The notations m_{reg} and m_{AK15} have been used interchangeably in the following sections.

The merged-jet approach is best suited for events where the charmed Higgs/Z boson candidate has a high Lorentz boost; in such cases, the angular separation between the two charm quarks is smaller and hence the entire Higgs/Z boson decay can be reconstructed within a cone of radius $\Delta R = 1.5$. In such events, the merged-jet approach has a very high signal reconstruction efficiency. However, such highly boosted events form a small part of the signal phase space (~5%). On the other hand, the resolved-jet approach can reconstruct cc events with well-separated c jets and hence has access to a larger part of the signal phase space. Therefore, the resolved-jet approach has been used at lower Lorentz boosts (or equivalently, lower p_T values) of the Higgs/Z boson candidates, while the merged-jet approach has been implemented at higher p_T values of the VHcc and VZcc searches.

It can be noted that the VWcq search in the merged-jet approach would require a dedicated cs tagger. Calibrating such a di-quark jet flavour tagger, in turn, would require a pure sample of cs jets in data which is difficult to obtain. Therefore, the VWcq search has been performed with two separately-reconstructed and individually-tagged AK4 jets for the entire p_T range of the reconstructed W boson.

References

1. CMS Collaboration Search for Higgs boson decay to a charm quark-antiquark pair in proton-proton collisions at $\sqrt{s}$ = 13 TeV. https://doi.org/10.48550/ARXIV.2205.05550. arXiv:2205.05550. Accepted for publication by PRL
2. CMS Collaboration (2020) A search for the standard model Higgs boson decaying to charm quarks. JHEP 03:131. https://doi.org/10.1007/JHEP03(2020)131. arXiv:1912.01662
3. Qu H, Gouskos L (2020) ParticleNet: jet tagging via particle clouds. Phys Rev D 101:056019. https://doi.org/10.1103/PhysRevD.101.056019. arXiv:1902.08570
4. CMS Collaboration (2020) Identification of highly Lorentz-boosted heavy particles using graph neural networks and new mass decorrelation techniques. CMS Detector Performance Note CMS-DP-2020-002. https://cds.cern.ch/record/2707946
5. CMS Collaboration (2022) Calibration of the mass-decorrelated ParticleNet tagger for boosted $b\bar{b}$ and $c\bar{c}$ jets using LHC Run 2 data. CMS Detector Performance Note CMS-DP-2022-005. https://cds.cern.ch/record/2805611
6. Larkoski AJ, Marzani S, Soyez G, Thaler J (2014) Soft drop. JHEP 05:146. https://doi.org/10.1007/JHEP05(2014)146. arXiv:1402.2657
7. CMS Collaboration (2021) Mass regression of highly-boosted jets using graph neural networks. CMS Detector Performance Note CMS-DP-2021-017. https://cds.cern.ch/record/2777006

Chapter 9
The VHcc, VZcc, and VWcq Analyses

This section presents the event selections used to perform the VHcc, VZcc, and VWcq analyses, along with short discussions on background estimation, corrections to simulations, and uncertainty analysis.

9.1 Data Samples

The Run-2 of the LHC began in 2015 and ended in 2018. During this period, the LHC delivered proton-proton collisions at a center-of-mass energy of 13 TeV. The amount of data collected in 2015 by the CMS experiment is an order of magnitude smaller than the amount of data collected in each of the years 2016, 2017, and 2018. Of the recorded data, some "runs" not passing the requisite criteria of data quality are not used in analyses. The other portion of the data that is certified as "good" has been used in the presented analyses. A summary of the corresponding integrated luminosities is provided in Table 9.1.

As discussed in Sect. 3.2.6, all recorded events are grouped into "primary datasets", that are not mutually exclusive, depending on which HLT path(s) the events have passed. For each of the analysis channels defined in Sect. 8.1, only events passing one or more HLT paths are analysed. Therefore, choosing only the primary dataset that contains all events passing the HLT path(s) used in a particular channel, ensures all triggered events necessary for that channel are processed, without the need to process the entire data collected by CMS. The datasets used thus depend on the channel being analysed. The list of primary datasets used for various channels across various data-taking years is presented in Table 9.2. The HLT paths that are used for each channel in each year are listed in Table 9.3.

S. Mondal, *Charming Decays of the Higgs, Z, and W Bosons*, Springer Theses,
https://doi.org/10.1007/978-3-031-84677-9_9

Table 9.1 Integrated luminosities of data collected during LHC Run-2 [1–4]

Year	Integrated luminosity (fb^{-1})			Uncertainty
	LHC delivered	CMS recorded	Used in analyses	
2015	4.31	3.86	0.00	2.7%
2016	41.6	38.2	36.3	2.5%
2017	49.8	45.0	41.5	2.3%
2018	67.9	63.7	59.7	2.5%
Run-2	164	151	138	–

Table 9.2 Names of the primary datasets used for various channels across various data-taking years. The definition of the electron datasets changed in the 2018 data-taking year. The reconstruction algorithms used for the datasets collected in 2016, 2017, and 2018 correspond to the CMSSW version 9_4_X, 9_4_X, and 10_2_X, respectively, and are generally referred to as "re-reco" or "pre-legacy" datasets

Channel		Dataset		
		2016	2017	2018
0L		`MET`	`MET`	`MET`
1L	μ	`SingleMuon`	`SingleMuon`	`SingleMuon`
	e	`SingleElectron`	`SingleElectron`	`EGamma`
2L	$\mu\mu$	`DoubleMuon`	`DoubleMuon`	`DoubleMuon`
	ee	`DoubleEG`	`DoubleEG`	`EGamma`

9.1.1 HEM15/16 Issue in 2018

In part of the dataset collected in 2018, namely from run 319077 onward, two sectors of HCAL Endcap Minus (HEM15 and HEM16) were not operational. This results in lower jet energy response in regions "$-1.57 < \phi < -0.87, -2.5 < \eta < -1.3$" and "$-1.57 < \phi < -0.87, -3.0 < \eta < -2.5$", as well as higher probabilities for jets to be misidentified as electrons. Therefore, in the analysis of the data collected in 2018, events are rejected if jets used for the Higgs/Z/W boson reconstruction lie in the affected regions. Additionally, events in the 1L channel are rejected if the selected electron lies in the affected regions. Overall, the effect of this issue is found to be small for the 1L and 2L channels and additional uncertainties are assessed according to the recipe provided by the JETMET POG.

However, this issue has a much more significant impact in the 0L channel. The lower jet energy response in the HEM failure regions leads to a momentum imbalance for dijet events, thus creating a large fake MET aligned with the HEM failure region, back-to-back with a high-p_T jet opposite to the HEM failure region. This mimics the signature of the Z($\nu\nu$)H signal events. Therefore, these mismeasured dijet events can pass the event selection of the 0L channel, leading to large discrepancy between simulation and the observed data. To veto these mismeasured dijet events, events

Table 9.3 The HLT paths used for various channels across various data-taking years. Wherever multiple HLT paths are indicated, an event is considered "passed" if it passes any one of the HLT paths

Channel		2016	2017	2018
0L		PFMET110_PFMHT110_IDTight	PFMET120_PFMHT120_IDTight	PFMET120_PFMHT120_IDTight
		PFMET120_PFMHT120_IDTight		
		PFMET170_NoiseCleaned	PFMET120_PFMHT120_IDTight_PFHT60	
		PFMET170_BeamHaloCleaned		
		PFMET170_HBHECleaned		
1L	μ	IsoMu24	IsoMu27	IsoMu24
		IsoTkMu24		
	e	Ele27_eta2p1_WPTight_Gsf	Ele32_WPTight_Gsf_L1DoubleEG	Ele32_WPTight_Gsf
2L	$\mu\mu$	Mu17_TrkIsoVVL_Mu8_TrkIsoVVL(_DZ)	Mu17_TrkIsoVVL_Mu8_TrkIsoVVL_DZ_Mass3p8	Mu17_TrkIsoVVL_Mu8_TrkIsoVVL_DZ_Mass3p8
		Mu17_TrkIsoVVL_TkMu8_TrkIsoVVL(_DZ)	Mu17_TrkIsoVVL_Mu8_TrkIsoVVL_DZ_Mass8	
	ee	Ele23_Ele12_CaloIdL_TrackIdL_IsoVL_DZ	Ele23_Ele12_CaloIdL_TrackIdL_IsoVL	Ele23_Ele12_CaloIdL_TrackIdL_IsoVL

with $-1.7 < \phi_{\mathrm{MET}} < -0.6$ for run $\geq$ 319077 in the data collected in 2018 are rejected, and then the MC simulation is scaled down accordingly by a factor of 0.3529 for events with $-1.7 < \phi_{\mathrm{MET}} < -0.6$. This leads to $\sim$ 10% loss of events in the 0L channel of the 2018 data-taking year.

9.2 Simulated Samples

The signal and background processes relevant to the analyses are simulated using standard Monte Carlo methods as discussed in Sect. 4.2. The quark-induced WH and ZH processes are generated at NLO accuracy in QCD using POWHEG (v2) extended with the MiNLO procedure [5, 6]. The gluon-induced ZH processes, on the other hand, are generated at LO accuracy with POWHEG. The signal samples for the VHcc search further involve a decay of the generated Higgs boson to a charm quark-antiquark pair. These signal samples along with their theoretical cross sections are listed in Table B.1 in Appendix B. Similarly, the VHbb process, which is a background, involves the Higgs boson decaying to a pair of b quarks, and have been listed in Table B.2 in Appendix B. All these samples have been generated with a Higgs boson mass of 125 GeV. The theoretical cross sections and branching ratios are provided by the LHC Higgs Cross Section Working Group [7].

Diboson samples form the signals for both the VZcc and VWcq analyses, and form a minor background in the VHcc analysis. In particular, only the events in simulated diboson qq samples that contain charm decays of the vector boson of interest at the generator level are considered to be signal. All other events in the qq samples, as well as other non-qq samples, are considered backgrounds. Diboson samples are simulated using either POWHEG and PYTHIA at LO, or MADGRAPH5_AMC@NLO at NLO. Their cross sections are normalised to those calculated at NLO for WZ and ZZ [8], and NNLO for WW [9]. Some samples have different cross sections for datasets corresponding to the 2016 and 2017–18 data-taking conditions owing to different underlying event tunes and PDFs. The diboson samples and their cross sections used along with the different datasets are listed in Tables B.3 and B.4 in Appendix B.

The V+jets samples (including DY+jets, W+jets, and Z($\nu\nu$)+jets) form the major backgrounds of all three analyses. They are generated at NLO precision with MADGRAPH5_AMC@NLO, with FXFX jet matching. Their cross sections have been rescaled to NNLO accuracy [10]. To optimise the available statistics of simulated samples all over the V p_{T} spectrum, samples are generated with various ranges of V p_{T} at the generator level. Furthermore, in datasets simulated using 2017 and 2018 data-taking conditions, V+jet samples have been split by the number of additional partons at the generator level. The masses of the Z bosons simulated in the DY and Z($\nu\nu$) samples are 50 GeV and above. The V+jets samples and their cross sections used along with the different datasets are listed in Tables B.6 and B.7 in Appendix B.

The $\mathrm{t\bar{t}}$ and Single Top (ST) processes are backgrounds in all analyses. The ME generation of the $\mathrm{t\bar{t}}$ simulation is performed with POWHEG (V2) at next-to-leading order (NLO) accuracy in QCD, and its cross section is scaled to a theoretical prediction at NNLO in QCD, including resummation of next-to-next-to-leading logarithmic soft-gluon terms. The single top quark production in the s-channel is also simulated using MADGRAPH5_AMC@NLO in the four-flavour scheme (four lightest quarks as the initial-state partons) in datasets corresponding to the 2016 data-taking conditions. The single top quark production in the t-channel in all data-taking years, and in the s-channel in datasets corresponding to the 2017 and 2018 data-taking conditions, is simulated using POWHEG also in the four-flavour scheme. The tW channel is simulated using POWHEG in the five-flavour scheme [11, 12], with its cross section normalised to the NLO calculations [13]. The $t\bar{t}$ and ST samples along with their cross sections have been listed in Table B.5 in Appendix B.

9.3 Event Reconstruction

Event reconstruction refers to the process of the reconstructing intermediate particles that are expected in signal events using reconstructed final state objects. As discussed in Sect. 8.1, events with a leptonically-decaying vector boson along with two jets that are likely to be charm jets from a vector boson decay, are selected for the so-called Signal Regions (SRs) of the three analyses. Depending on the channel under consideration (0L, 1L, or 2L), different numbers of charged leptons and different magnitudes of MET are required to describe the vector boson decay. The following subsections describe the reconstruction of the leptonically-decaying vector boson and the Higgs/Z/W boson candidate decaying into charm quark(s).

9.3.1 Leptonically-Decaying Vector Boson Reconstruction

Depending on the channel, the leptonically-decaying vector boson (V), either W or Z, is reconstructed using either a pair of charged leptons, a combination of charged lepton and MET, or MET alone. Unless specifically mentioned, the vector boson reconstruction strategy is the same across the VHcc, VZcc, and VWcq analyses. The following requirements are used for each channel:

- **2L** ($Z \rightarrow \mu^+\mu^-/e^+e^-$) **channel**: Two muons (electrons) of opposite charge satisfying the criteria for "isolated, prompt muons (electrons) in 2L channel", as defined in Sect. 5.1 (5.2), are required in each event. The Z boson candidate is reconstructed by adding the four momenta of the two muons (electrons). Only events in which the invariant mass of the reconstructed Z boson candidate (m_{ll}) lies between 75 and 150 GeV are considered. In the VHcc and VZcc analyses, this channel is further divided into two categories depending on the p_T of the reconstructed Z (p_T^Z): **(i)** 2-lepton-low (2LL), when 60 GeV $< p_T^Z <$ 150 GeV, and **(ii)** 2-lepton-high (2LH), when $p_T^Z >$ 150 GeV.

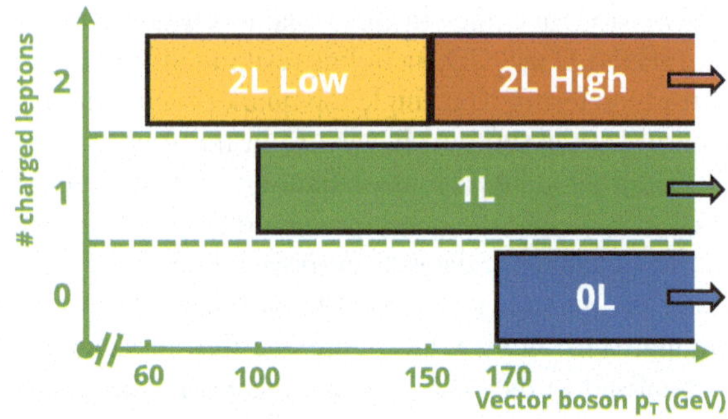

Fig. 9.1 A graphical representation of the three channels with the vector boson p_T thresholds used in the VHcc and VZcc analyses (not to scale). The 2L channel is not split in the VWcq analysis

- **1L** (W $\rightarrow \mu\nu/e\nu$) **channel**: One muon (electron) satisfying the criteria for "isolated, prompt muons (electrons) in 1L channel", as defined in Sect. 5.1 (5.2), is required in each event. It is then combined with the MET of the event to define the p_T of the W boson candidate,

$$p_T^W = \sqrt{\left(\text{MET}_x + p_x^l\right)^2 + \left(\text{MET}_y + p_y^l\right)^2}, \tag{9.1}$$

  where MET_x (MET_y) and p_x^l (p_y^l) are the x (y) components of the MET and lepton, respectively. Only events with $p_T^W > 100$ GeV are considered. If the event contains one or more additional muons (electrons) satisfying the criteria of "veto muons (electrons) in 0L and 1L channels", defined in Sect. 5.1 (5.2), the event is discarded.
- **0L** (Z $\rightarrow \nu\nu$) **channel**: Events with MET > 170 GeV are considered. If the event contains one or more muons (electrons) satisfying the criteria of "veto muons (electrons) in 0L and 1L channels", defined in Sect. 5.1 (5.2), the event is discarded. The MET represents the p_T of the reconstructed Z boson candidate.

A graphical representation of the three channels with the vector boson p_T thresholds is presented in Fig. 9.1. The mass (applicable in 2L channel only) and p_T of the reconstructed vector boson are denoted by m_V and $p_{T,V}$ respectively.

9.3.2 Hadronically-Decaying Boson Reconstruction

This subsection discusses the reconstruction of the Higgs boson, Z boson, and W boson that decay into charm quark(s) in the VHcc, VZcc, and VWcq analyses, respectively.

All selected events are required to have at least 2 well-reconstructed AK4 jets satisfying tight `JetID` and loose `PileupJetID` (for jets with $p_T < 50$ GeV), as discussed in Sect. 5.3. The selected jets are then sorted by their DeepJet CvsL scores. The jets with the highest (j_1) and second-highest (j_2) scores are considered to be the

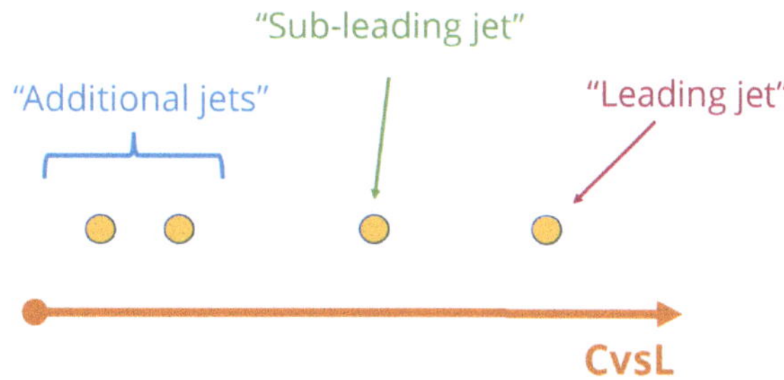

Fig. 9.2 A graphical representation of charm jet candidate selection in a hypothetical event. The leading and subleading jets are selected as the decay products of the boson (Higgs/Z/W). Additional jets may or may not be present

two charm jet candidates for the VHcc and VZcc analyses, and charm and strange jet candidates, respectively, in the VWcq analysis. A graphical representation of the same is shown in Fig. 9.2. Sorting by other parameters of jets, such as jet p_T and jet CvsB score, was also tested, but sorting by CvsL score led to identification of correct Higgs boson decay products with the highest efficiency.

An alternate approach to perform the jet-pair selection using a BDT was also tested, with the CvsL, CvsB, p_T of the two jets, alongside angular separations between the jets, as inputs to the BDT, using simulated VHcc samples. However, using the BDT introduced a so-called "mass sculpting" in the background events, which means the BDT preferentially selected jet pairs with invariant masses (m_{jj}) close to the Higgs boson mass, even in background events. This indicates that the BDT learns information about the Higgs boson mass by exploiting correlations among the provided inputs. This is detrimental to the signal-background separation to be performed in the next step by making use of the m_{jj} information of the signal and background events. Therefore, this approach was finally not implemented, even though using the BDT expectedly improves the efficiency of identifying the correct Higgs-boson-decay jet pair. However, this approach can be improved in the future by training the BDT with simulated pseudo-Higgs boson samples with Higgs boson masses varying in a flat spectrum.

A comparison of the jet-pair selection efficiencies using the different approaches is presented in Table 9.4. The four momenta of the selected jets are then added to obtain that of the Higgs/Z/W boson candidate. The mass and p_T of the Higgs/Z/W boson candidate are denoted by m_{jj} and $p_{T,jj}$, respectively.

9.3.3 *Top Quark Candidate in 1L Channel*

Semileptonic $t\bar{t}$ events form a major background in the 1L channel due to a final state similar to that of VHcc/VZcc/VWcq signal events with 1 charged lepton. An additional handle to discriminate the signal against $t\bar{t}$ events is introduced by explicitly reconstructing a top quark candidate in this channel from the final state decay products. The mass and kinematics of this reconstructed top candidate is expected

Table 9.4 The efficiency of selecting the correct jet pair in the VHcc search using various jet variables to sort jets. While using the BDT expectedly improves the efficiency of identifying the correct Higgs-boson-decay jet pair, its mass sculpting effect diminishes the signal-background discrimination in the next steps. Hence, CvsL-sorting of jets is used instead

Channel		Correct jet-pair selection efficiency (%)			
		p_T	CvsL	CvsB	BDT
0L		73.5	73.2	71.4	74.3
1L	μ	59.3	66.9	51.9	70.5
	e	58.6	66.5	51.2	70.2
2L	$\mu\mu$	59.8	66.6	50.1	70.9
	ee	59.7	66.5	50.8	70.8

to be different in $t\bar{t}$ events (a true top quark) than in signal events (a non-physical combination of observables).

The top quark candidate is reconstructed using the isolated prompt charged lepton selected in the 1L channel combined with a b jet candidate and a neutrino candidate. The b jet from the leptonic branch of $t\bar{t}$ is identified by using a loose b tagging criterion (a selection criterion on the DeepJet BvsC/L score) and selecting the jet closest to the charged lepton (in terms of ΔR) among all the jets that pass this b tagging criterion. The neutrino and the charged lepton in a true semileptonic $t\bar{t}$ event would arise from a W boson, and therefore the invariant mass of the two is expected be close to the W boson mass (m_W). Constraining the invariant mass of the two to $m_W = 80.38$ GeV, the following equation is obtained,

$$m_W^2 = (E_l + E_\nu)^2 - \left(p_{x,l} + p_{x,\nu}\right)^2 - \left(p_{y,l} + p_{y,\nu}\right)^2 - \left(p_{z,l} + p_{z,\nu}\right)^2, \qquad (9.2)$$

where E_l is the energy of the lepton, $p_{x,l}$, $p_{y,l}$, and $p_{z,l}$ ($p_{x,\nu}$, $p_{y,\nu}$, and $p_{z,\nu}$) are the x-, y-, and z-components, respectively, of the lepton (neutrino) momentum, and $E_\nu = \sqrt{p_{x,\nu}^2 + p_{y,\nu}^2 + p_{z,\nu}^2}$ is the energy of the neutrino. As $p_{x,\nu}$ and $p_{y,\nu}$ can be estimated from the x- and y-components of the MET, $p_{z,\nu}$ is the only unknown quantity in Eq. 9.2. The equation is quadratic in $p_{z,\nu}$ and can be solved to obtain two possible values for this quantity. The z-component of the neutrino momentum is defined as the lower of the two solutions for $p_{z,\nu}$. Thereafter, the four-momenta of the charged lepton, the b jet candidate, and the neutrino are combined to reconstruct a top candidate.

A number of kinematic variables related to the top candidate along with flavour tagging variables associated with the b jet candidate are used in the signal extraction in the 1L channel, as described in Sect. 10.1. The distribution of the mass of the reconstructed top quark candidate is shown in Fig. 9.3.

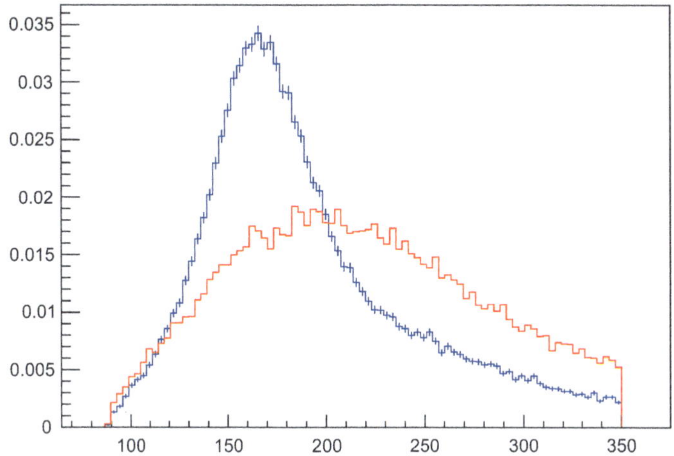

Fig. 9.3 The normalised distribution of the mass of the reconstructed top quark candidate in the simulated VHcc signal (red) and $t\bar{t}$ background (blue), plotted upto 350 GeV. The simulated samples correspond to the 2018 data-taking conditions. The blue plot exhibits a sharper peak around the top quark mass (~172 GeV) compared to the red plot

9.3.4 Other Event Parameters

Some additional event-level variables are defined to select signal events and reject background. They are defined as follows:

- **Additional jets,** N_{aj}: The number of additional jets is defined as the number of jets with $p_{\mathrm{T}} > 30\,\mathrm{GeV}$ and $|\eta| < 2.5$, passing the tight `JetID` and loose `PileupJetID`, *in excess of* the two selected jets and the FSR-recovered jets (cf. Sect. 9.4.1).
- **Additional leptons,** N_{al}: The number of charged leptons in the event satisfying the criteria of "veto muons/electrons in 0L and 1L channels" (cf. Sects. 5.1 and 5.2), in addition to the one(s) selected for vector boson reconstruction.
- $\Delta\phi_{\mathrm{MET},l}$: The difference in the azimuthal angle (in radians) between the MET and the selected lepton in the 1L channel.
- $\Delta\phi_{\mathrm{V},jj}$: The difference in the azimuthal angle (in radians) between the reconstructed leptonically-decaying vector boson and the reconstructed Higgs/Z/W boson.
- **MET significance**: It is defined as the ratio of the MET to the square-root of the scalar sum of the hadronic transverse energy of all jets with $p_{\mathrm{T}} > 30\,\mathrm{GeV}$.

9.4 Refinement of Dijet Invariant Mass

A number of novel advanced analysis techniques have been implemented to further refine the dijet invariant mass (m_{jj}). This mass is of particular interest as it helps in distinguishing the resonant Higgs/Z/W boson signal from the backgrounds, and from each other. Hence, any improvements to its accuracy and/or resolution is expected to improve the signal sensitivity of the analyses. The following subsections discuss the techniques used in the presented analyses.

9.4.1 FSR Jet Recovery

In addition to the two jets selected as the Higgs/Z/W boson decay candidates, an FSR jet recovery is performed whenever a suitable candidate is present. A final-state radiation of a soft gluon is a common process occurring during jet hadronisation. The angle of the emitted gluon can be large enough to lie outside the clustering cone of the main jet. Additional jets arising from FSRs can be added back to the main jet, improving the measurement of its p_{T}. Such a procedure is implemented in the analyses by adding additional jets with $p_{\mathrm{T}} > 20$ GeV and $|\eta| < 3.0$ which are reconstructed within $\Delta R < 0.8$ of either of the two selected jets. Recovering the FSR jets improves the mass resolution of the Higgs/Z/W boson by a few percent, without sculpting the background shape. A comparison of the dijet mass and mass resolution before and after FSR-recovery in simulated VHcc events is presented in Table 9.5.

Table 9.5 Table showing the relative improvement due to the FSR recovery in the dijet invariant mass distribution. For each channel, the mean (μ) and standard deviation (σ) in GeVs resulting from a double-sided Crystal-Ball fit on the dijet mass are shown. Improvement in accuracy is defined as the relative change in the quantity $|\mu - 125|$ after FSR recovery, while improvement in resolution is defined as the negative of the relative change in σ/μ

Channel	Without FSR			With FSR			Improvement (%)	
	μ	σ	σ/μ	μ	σ	σ/μ	Resolution	Accuracy
0L	125.69	13.1	0.105	125.73	12.9	0.103	+1.9	–5.8
1L	122.95	14.1	0.115	123.75	14.1	0.114	+0.9	+39.0
2LL	116.87	17.3	0.148	118.73	16.2	0.136	+8.1	+22.9
2LH	121.30	14.3	0.118	122.66	14.1	0.115	+2.5	+36.8

9.4.2 Charm Jet Energy Regression

Jet energy corrections, as explained in Sect. 5.3, are implemented to adjust energy measurements of jets to reflect the kinematics of the originating particle more accurately. However, these energy corrections have a dependence on the flavour of the jets. Energies of bottom and charm jets, for example, are usually mismeasured to a greater extent compared to light-flavour jets, due to various factors, such as, the jet cone not containing all products of the heavy hadron decay, and presence of unmeasured neutrinos arising from a semileptonic decay of the b or c hadron. The latter happens in about $\sim 20\%$ of b jets and $\sim 10\%$ of c jets. The phase spaces where the JECs are measured centrally by the CMS JetMET POG are usually dominated by light-flavour jets, and hence there is room for improving the energy measurements in analyses dominated by b and c jets.

In this context, a dedicated method [14] to estimate the energy and resolution of b jets was developed for use in the VHbb analysis [15]. The method used a dedicated multivariate DNN-based regression algorithm to obtain point and dispersion estimates for the energies of b jets. A similar algorithm has been adopted for c jets to estimate the energy and per-jet energy resolution of c jets using a DNN that uses composition and shape information of simulated c jets, and the properties of reconstructed secondary vertices associated with the jet.

The application of c jet energy regression to the selected leading and subleading jets of the analyses results in an improvement in the m_{jj} resolution of about 20% in simulated VHcc events. Figure 9.4 shows an example of the improvement in the 2L channel of the VHcc analysis with simulations corresponding to the 2018 data-taking conditions. A similar improvement is expected in the Z and W boson mass in the VZcc and VWcq analyses.

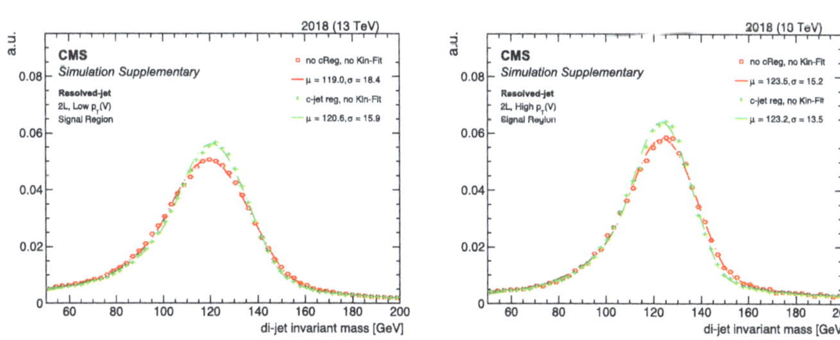

Fig. 9.4 Invariant mass distribution (m_{jj}) of the reconstructed Higgs boson candidate obtained in the 2LL (left) and 2LH (right) channels in simulations corresponding to the 2018 data-taking conditions. Distributions using jets with and without c jet energy regression applied are shown, together with the results of a double-sided Crystal-Ball fit

9.4.3 Kinematic Fit in 2L Channels

Particularly in the 2L channels, the resolution of the p_T of the selected jets can be additionally improved with a kinematic fit (kinfit) procedure. The principle behind this method is that the total momenta of all the final-state reconstructed objects must be zero in the transverse plane, as there is no "true" missing energy in the VHcc final state in the 2L channel. This constraint can be used in a maximum-likelihood fit, where the momenta of the jets are allowed to vary within their uncertainties obtained from the energy measurement.

In addition to p_T conservation, one can also exploit the fact that leptons in the final state are the decay products of a Z boson, which has a well-known mass and natural width. This constraint is also used in the kinematic fit. Although this alone does not have a great impact on the m_{jj} distribution, when combined with the momentum conservation it results in an improved measurement of the jet momenta and related variables. More details about the kinematic fit method implemented in the VHcc analysis can be found in Ref. [16].

The resolution of the Higgs boson candidate mass in VHcc signal events improves significantly after application of the kinematic fit procedure, as seen in Fig. 9.5. The improvement is higher in the 2LH channel compared to that in 2LL channel. The effect of kinfit alone on the m_{jj} resolution amounts to 36% improvement for the 2LH and 14% for the 2LL channel. The c jet energy regression and kinfit used together result in a 40% improvement in the m_{jj} resolution for the 2LH channel and 22% for the 2LL channel.

The kinfit has not been applied in the VWcq search as the W mass peak is expected to be far away from the Higgs boson peak, for which the kinfit algorithm was set up.

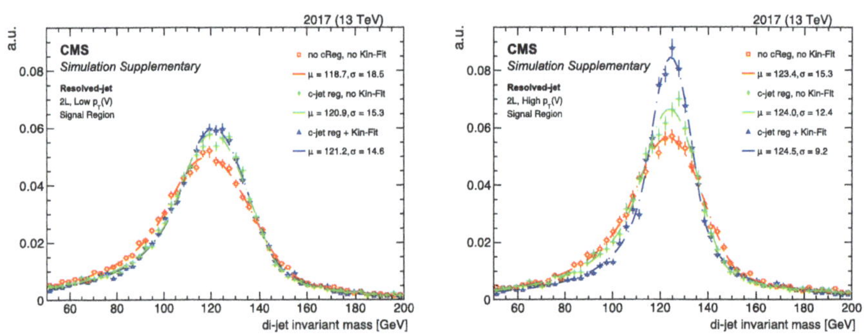

Fig. 9.5 Invariant mass distribution of the reconstructed Higgs boson candidate obtained from a VHcc signal sample, simulated using the 2017 data-taking conditions, in the 2LL (left) and 2LH (right) channels, together with the results of a double-sided Crystal-Ball fit. The red histograms show the distribution without applying c jet energy regression or kinematic fit. The yellow histograms are obtained after applying kinematic fit, using jets without regression. The green histograms are with regression, but without kinematic fit. Finally, the blue histograms show the distribution when both the regression and the kinematic fit are applied

9.5 Event Selection Criteria

Applying selection criteria ("cuts") for event selection refers to the process of identifying a subset of the pp collision events recorded by the CMS experiment in a way such that the subset retains as many signal events as possible while suppressing background processes. The criteria are usually chosen in a physics-motivated way and by making use of simulated events. The general idea in the presented analyses is to require a well-reconstructed leptonically-decaying vector boson, and a well-reconstructed Higgs/W/Z boson, each with sufficient Lorentz boost, to get rid of the main backgrounds. Moreover, as the two bosons are expected to recoil against each other, they are required to be well-separated in the ϕ axis.

In all three analyses, the leading jet (leading in CvsL score) is required to pass a charm jet selection criteria. This criteria has been optimised to approximately maximise the signal significance, by scanning the Asimov significance[1] of the VHcc signal at various pairs of (CvsL,CvsB) values of the leading jet. The final selection criteria chosen for the leading jet is CvsL $>$ 0.225 and CvsB $>$ 0.400. This c tagging selection corresponds to approximately 30–40% charm jet efficiency, with a 4% light-flavour jet mistag rate and 16% b jet mistag rate, for the leading jet. The b and light-flavour mistag rates are lower in the 2017–18 data-taking conditions compared to those in 2016 due to the tracker upgrade. The c tagging ROC contours along with the working point used for the leading jet for the datasets corresponding to the 2016 and 2018 data-taking conditions are shown in Fig. 9.6. A similar ROC contour for the 2017 data-taking condition was shown in Fig. 7.9 (right) in the context of charm tagger calibration.

The exact selection criteria on the various reconstructed objects and event parameters depend on the channel and are listed in Table 9.6. Application of these selection criteria on the entire dataset yields the so-called Signal Region (SR) of the analyses. SRs are defined for each channel, in each year of data-taking.

9.5.1 *Extra Selection Criteria in the VWcq Analysis*

In addition to the above mentioned selections, certain additional steps are performed in the VWcq analysis. One of them is the implementation of a BDT that has been discussed in Sect. 8.2 in the context of correlated charm-strange tagging. The BDT, as discussed, is trained to discriminate simulated VWcq signal events against simulated VZcc events, which form a background for this analysis. The BDT makes use of the CvsL and CvsB scores of the 2 selected jets as inputs. A separate BDT is trained per channel, per data-taking year. Finally, a selection criterion on this BDT is applied to select only VWcq-like events, rejecting a majority of VZcc events. This makes

[1] The Asimov significance can be approximated using $S/\sqrt{B}$, where S is the signal yield and B is the background yield.

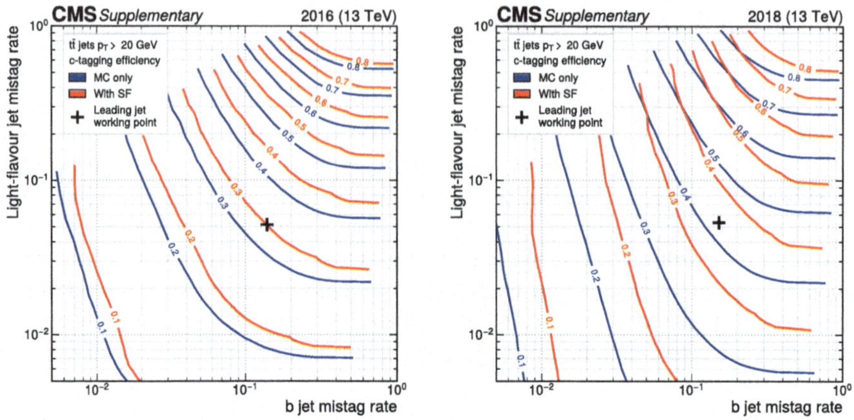

Fig. 9.6 The ROC contours showing c tagging efficiencies as functions of b and light-flavour jet misidentification rates, for the DeepJet algorithms for simulated jets (the blue lines) and the estimation of the same for jets in data (the red lines), for the datasets corresponding to the 2016 (left) and 2018 (right) data-taking conditions. Each line represents points in the plane that correspond to a fixed value of the c tagging efficiency, which is shown as a number at the centre of each line. The black cross denotes the working point used for the leading jet selected in the three analyses

Table 9.6 The various selection criteria used across the three channels. All p_Ts and masses are in GeV and all angles are in radians. Variables are defined in Sect. 9.3

Variable	0L	1L	2LL	2LH
$p_{T,V}$	> 170	> 100	[60,150]	> 150
m_V (m_{ll})	–	–	[75,105]	
Lepton p_T	–	>25 (μ), >30 (e),	> 20	
Leading jet p_T	> 60	> 25	> 20	
Subleading jet p_T	> 35	> 25	> 20	
$p_{T,jj}$	> 120	> 100	–	
m_{jj}	[50,200] (VHcc, VZcc) or [50,120] (VWcq)			
Leading jet CvsL	> 0.225			
Leading jet CvsB	> 0.400			
N_{aj}	–	< 2	–	
N_{al}	= 0	= 0	–	
MET	> 170	–	–	
MET significance	–	> 4.0	–	
$\Delta\phi_{V,jj}$	> 2.0	> 2.5	> 2.5	
$\Delta\phi_{MET,l}$	–	< 2.0	–	

VZcc contribution in the SR negligible and discards the need to perform a data-driven estimation of this background.

Another step taken for the VWcq analysis is the introduction of an additional low-CvsL signal region in the 0L channel. This additional SR is defined by inverting the CvsL criterion mentioned in Table 9.6, and applying all the other criteria as mentioned. It has been tested that using this additional SR in conjunction with the high-CvsL SR recovers more signal events compared to extracting signal from the high-CvsL SR alone, and the difference in expected significance of the VWcq signal among the two approaches is significant in the 0L channel. While the same additional SR has been defined in 1L and 2L channels too, they have not been used in the final signal extraction fit owing to an insignificant increase in expected significance.

9.6 Data-Driven Background Estimation

Background events which remain in the SRs after application of the selection criteria (irreducible backgrounds) need to be modelled accurately to perform a hypothesis testing of the S+B hypothesis against the B-only hypothesis. Since simulations are often prone to mismodellings and theoretical cross sections of processes may be inaccurate, a data-driven approach is adopted to estimate the event yields of the major backgrounds in the SR. These corrections need to be specific to the selection phase space and therefore any selection of events used to estimate these backgrounds needs to be as close to the SR as possible. This leads to the definition of so-called Control Regions (CRs), each of which is enriched with a specific background process so that correction to that process may be derived accurately using this CR.

9.6.1 Background Definitions

The major backgrounds in this analysis are V+jets and $t\bar{t}$. The relative flavour compositions of the selected jets in the V+jet selections are also prone to mismodellings in the simulation and hence are estimated separately in separate CRs. Therefore, the simulated V+jets processes are further split by the flavour of the selected jets, where the flavour definition of a jet is the same as used in jet flavour tagging algorithms (cf. Sect. 6). The definitions of V+jets sample split by flavour are as follows:

- **V+b jet**: At least one of the jets in the V+jets event is a b jet.
- **V+c jet**: At least one of the jets in the V+jets event is a c jet, but none of the jets are b jets.
- **V+light jet**: V+jets event that does not come under any of the above categories.

In this way, any V+jet event comes under exactly one of the above categories.

In the VWcq search, simulated $t\bar{t}$ events are categorised as well, but only for informational purposes; each category is not estimated separately from data. The definition of this splitting is as follows:

- **$t\bar{t}$(cs)**: Semileptonic $t\bar{t}$ events with at least one jet containing a c hadron arising from a generated W boson.
- **$t\bar{t}$(ud)**: Semileptonic $t\bar{t}$ events which does not follow the above criterion.
- **$t\bar{t}$**: Dileptonic and fully hadronic $t\bar{t}$ events.

9.6.2 Control Region Definitions

The motivation to model four distinct major backgrounds in a data-driven way, separately for each channel and each year of data-taking, leads to the definition of four CRs in each channel, in each year of data-taking: V+light jet CR, V+b jet CR, V+c jet CR, and $t\bar{t}$ CR, each of which aims to contain a significant number of events of the respective process. A brief description of the CR selections is as follows:

- **V+light jet (V+LF) CR**: This CR is defined by inverting the CvsL criterion mentioned in Table 9.6. This enriches the CR in V+udsg jet events. In the VWcq analysis, a further requirement of $m_{jj} > 120\,\text{GeV}$ is imposed to keep this CR orthogonal to the low-CvsL SR.
- **V+b jet (V+HF) CR**: This CR is defined by inverting the CvsB criterion mentioned in Table 9.6. In the 2L channel, a stricter criterion on the m_{ll} value (85 GeV $< m_{ll} <$ 97 GeV) is additionally required. This enriches this CR in V+b jet events.
- **V+c jet (V+CC) CR**: This CR is defined by inverting the m_{jj} criterion, that is, by requiring $10\,\text{GeV} < m_{jj} < 50\,\text{GeV}$ *or* $200\,\text{GeV} < m_{jj} < 250\,\text{GeV}$. The second criterion is changed to $120\,\text{GeV} < m_{jj} < 250\,\text{GeV}$ in the VWcq analysis. In the 2L channel, a stricter criterion on the m_{ll} value (85 GeV $< m_{ll} <$ 97 GeV) is additionally required. The flavour composition of this CR is similar to the SR, but it does not include the W/Z/Higgs boson resonances, and therefore is suited to estimate non-resonant charm jet productions.
- **$t\bar{t}$ (TT) CR**: The definition of the $t\bar{t}$ CR is dependent on the channel. Firstly, the CvsB criterion mentioned in Table 9.6 is inverted (in the same fashion as in the V+HF CR), and the $\Delta\phi_{V,jj}$ criterion is relaxed. However, in addition, **(i)** in the 2L channel, the m_{ll} and N_{aj} criteria are inverted, that is, $m_{ll}\,(\text{GeV}) \notin [75, 120]$, $m_{ll} > 10\,\text{GeV}$, and $N_{\text{aj}} \geq 2$ requirements are imposed, **(ii)** in the 1L channel, at least two additional jets are required ($\mathbf{N_{aj}} \geq 2$) and MET is required to be less than 170 GeV to keep this CR orthogonal to the TT CR of the 0L channel, and **(iii)** in the 0L channel, at least one additional lepton and at least two additional jets are required ($N_{\text{aj}} \geq 2,\ N_{\text{al}} \geq 1$).

The m_{jj} requirements are relaxed for all CRs except for the V+CC CR, to increase the event yields of the background processes. Graphical representations of the SR and

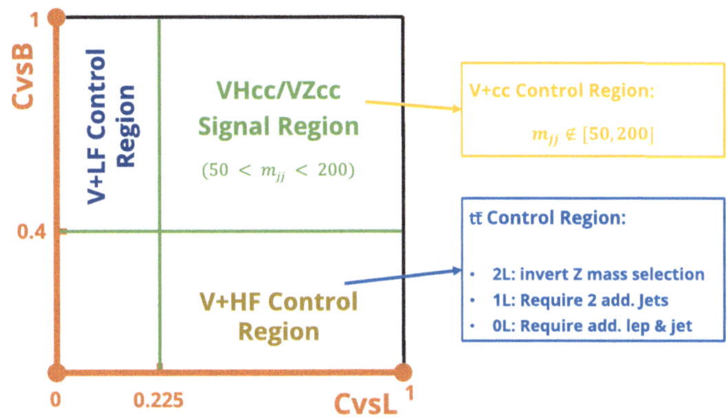

Fig. 9.7 A graphical representation of the SR and the CRs for each channel of the VHcc and VZcc analyses in the CvsL–CvsB 2D plane (not to scale). Detailed selection criteria explained in text

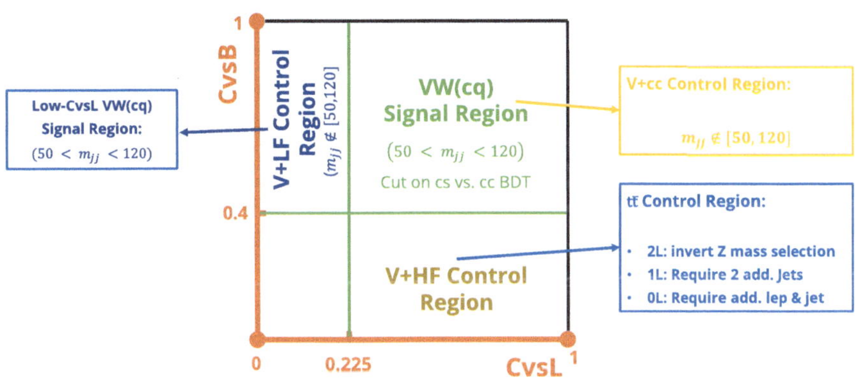

Fig. 9.8 A graphical representation of the SR and the CRs for each channel of the VWcq analysis in the CvsL–CvsB 2D plane (not to scale). Detailed selection criteria explained in text. The low-CvsL SR has been used only in the 0L fit

CRs in the CvsL–CvsB plane are shown in Figs. 9.7 and 9.8 for the VHcc/VZcc and VWcq analyses, respectively. Some distributions in the CRs of the VHcc analysis with the datasets corresponding to the 2017 data-taking conditions are shown in Fig. 9.9.

The CRs thus defined, are used simultaneously with the SRs in the binned maximum likelihood fit, allowing the yield of each of the defined major-backgrounds (cf. Sect. 9.6.1) to float freely. The factor, by which each freely-floating simulated process is scaled, is known as the rate parameter (RP) or normalisation factor for that process. In this context, it can be noted that the fit variables were seen to have similar distributions for V+b jet and V+c jet events in the 1L channel. This is why a common RP was used for these two processes specifically in the 1L channel. To account for this approximation, an additional 50% log-normal uncertainty was assigned to the V+b jet template in this channel.

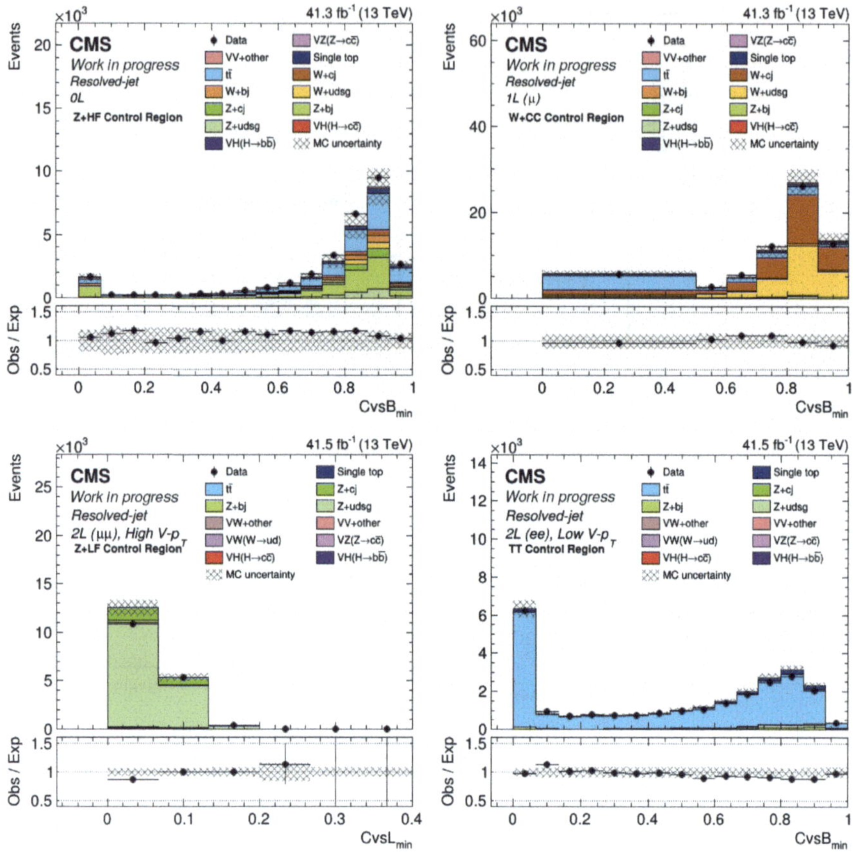

Fig. 9.9 Pre-fit charm tagger distributions of the subleading jet in the 0L HF CR (upper-left), 1L(μ) CC CR (upper-right), 2LL($\mu\mu$) LF CR (lower-left), and 2LH(ee) TT CR (lower-right) of the VHcc analysis with the datasets corresponding to the 2017 data-taking conditions. The bottom panel in each plot shows the ratio of data to the total simulation, with the uncertainty in simulation indicated by gray hatching. The bins in the LF and TT CRs are merged before the fit is performed

9.7 Corrections to Simulation

Simulations are often prone to mismodellings. These limitations can potentially manifest themselves in the distributions of the variables used in the final fit. Therefore, it is necessary to resolve any discrepancies observed between the simulation and data in the relevant phase space. In the same spirit as in the discussion in Sect. 6.3, these discrepancies can arise from a number of mismodellings in the simulation chain—starting from the MC generators and parton showering to the detector simulation and reconstruction. The effects of these discrepancies are usually adjusted for by reweighting simulated events with scale factors (SFs). These corrections are derived by comparing data and simulation in independent selections. An example of such a correction is the charm tagging SF, which has been discussed in detail in Sect. 7 and Ref. [17]. Other major corrections are discussed in the following subsections. These corrections are generally considered independent and multiplied on a per-event basis to assign a net weight to every simulated event.

9.7.1 Standard Corrections

This subsection lists the corrections that are applicable to a broad phase space of selections and are usually applied to all analyses of pp collision data in CMS. These are usually derived centrally by various POGs of the CMS collaboration for use across the collaboration.

Number of pileup vertices: The effect of pileup is simulated using PYTHIA in each simulated pp collision event by including the final-state products of a number of minimum-bias events within the event. The number of pileup vertices per simulated event is sampled from a Poisson distribution, which is modelled based on the observed number of pileup vertices in a limited amount of data. However, this number depends on the instantaneous luminosity delivered by the LHC and other factors and varies over the period of data collection. Therefore, correction weights are derived as a function of the number of interaction vertices, comparing its distribution in data to that in simulation.

Trigger efficiencies: Trigger paths are usually designed in a way to select events containing at least one electron, muon, or MET above a certain p_T threshold. While ideally all events above (below) this threshold should be accepted (rejected) by the triggers, practically there is a "turn-on" of the trigger acceptance around this threshold. This means some events with an object p_T lower (higher) than the threshold are also accepted (rejected) and the frequency of such occurrences decreases at p_Ts further away from the threshold. This effect is accounted for by applying the efficiency of the trigger as an event weight. The same efficiencies that were derived and used in the CMS VHbb analysis [15] have been used for the analyses presented in this thesis. For leptonic triggers, the efficiencies are evaluated as a function of lepton p_T and η exploiting DY ($Z/\gamma^* \rightarrow \mu^+\mu^-/e^+e^-$) events selected in data. For MET triggers, the efficiency is evaluated as a function of MET using W+jets and $\text{t}\bar{\text{t}}$ events, collected using a single lepton trigger and containing a lepton with p_T well above the p_T-threshold of the single lepton trigger.

Lepton efficiencies: Efficiencies of the selected electrons and muons are generally different in data and simulation after all the object selection criteria (Sects. 5.1 and 5.2) are applied. The differences are measured centrally by the EGAMMA and Muon POGs of the CMS Collaboration, comparing simulation and data in DY event selections. The corrections thus derived are factorised into reco and ID SFs for electrons, corresponding to the data-to-simulation efficiency ratios in the electron reconstruction and MVA-based electron ID, respectively. For muons, the SFs are factorised into reco, ID, and isolation SFs. While the reco and ID SFs derived centrally by the Muon POG have been used in the presented analyses, the isolation SFs for a tighter working point (< 0.06) derived for the CMS VHbb analysis [15] have been used in certain selections, in conjunction with the centrally-derived looser isolation criteria in others.

Jet energy corrections: As already discussed in Sect. 5.3, jet energies are subjected to corrections in both data and simulations. Simulations need to be additionally smeared with JER corrections to reflect the correct energy resolution in data. However, the usual JER corrections derived by the JetMET POG in CMS are not used in this analysis.

9.7.2 Charm Tagger Reshaping

A full reshaping of the DeepJet CvsL and CvsB scores of all selected jets is performed using the SFs derived, as described in Sect. 7. The set of SFs are separately derived for the datasets corresponding to the 2016, 2017, and 2018 data-taking conditions. Each SF value, in turn, is a function of a simulated jet's flavour, CvsL, and CvsB value. The event weight assigned to each event is the product of the SF values of all selected jets in the event. Thus, the c tagging weight of an event is given by

$$w_{\mathrm{c\ tag}} = \prod_{i \in \mathrm{jets}} \mathrm{SF}_i(\mathrm{flav}_i, \mathrm{CvsL}_i, \mathrm{CvsB}_i), \tag{9.3}$$

where flav stands for the flavour of the jet.

9.7.3 Smearing of Regressed Charm Jet Energy

Simulated jets are often reconstructed with a better energy resolution than what is observed in data. Hence, the energies of the simulated jets are smeared so that they match the resolution in data. However, since a dedicated charm energy regression (cf. Sect. 9.4.2) has been used on the selected jet candidates in the presented analyses, the usual JER corrections derived by the JetMET POG in CMS are not applicable. Instead, these analyses make use of the same smearing that was developed for regressed b jet energies [14] in the VHbb analysis [15]. The smearing is applied only to the VH and diboson samples as they are the only samples containing resonances that can be altered by the effect of the smearing.

The energy resolutions of jets can generally be measured in events where a jet recoils off a Z boson decaying into charged leptons. The Z boson and the jet are expected to have opposite momenta in the transverse plane. As the Z boson's p_{T} is well-reconstructed from the final-state leptons, the ratio of the jet's reconstructed p_{T}

to that of the Z boson's allows an estimate of the jet p_T resolution. The method yields a smearing of 5% for samples simulated using 2017 data-taking conditions, and 15% for those with 2018 data-taking conditions. No smearing was found necessary for those with 2016 data-taking conditions.

9.7.4 Z Boson p_T Reweighting in Z+jets

The p_T spectra of the reconstructed Z bosons in simulations at NLO in QCD were found to have a disagreement with that in data, at low values of the p_T ($<\sim$ 150 GeV). This effect was observed only in the 2LL channel (or 2L channel of the VWcq analysis), as the 0L channel uses only highly boosted Z boson candidates ($>$ 170 GeV). Corrections were derived by performing a linear fit to the data-simulation ratio of the distribution of the vector boson p_T in the 2L LF CR, as this region is particularly enriched in DY+jet events. Separate corrections for data-taking conditions in 2016, 2017, and 2018 were derived. An illustration of the effect of this correction is shown in Fig. 9.10 (left).

9.7.5 Jet-Angular-Separation Dependent Reweighting

A disagreement between data and simulations of V+jets events at low angular separation of the two jets (ΔR_{jj}) was observed in the CRs. This effect was found more prominent in the NLO samples used in the analyses compared to LO samples. In particular, the disagreement is visible in regions with $\Delta R_{jj} < 1.0$. It can be attributed to the mismodelling in gluon splitting events with small opening angles. An example of the disagreement is shown in Fig. 9.10 (right).

The disagreement is mitigated by a dedicated reweighting of simulated V+jets events containing two jets with $\Delta R_{jj} < 1.0$. The reweightings are derived from the LF CRs of the 2L, 1L, and 0L channels separately, by fitting 1st, 1st, and 3rd order polynomials respectively to the data-simulation ratios of the V+jet events after other minor backgrounds are subtracted off. Different corrections were derived for the three years. Additionally, a small flavour dependence of these corrections was observed at very low values of ΔR_{jj} (< 0.55). Inflated uncertainties are used for events with $0.4 < \Delta R_{jj} < 0.55$ containing heavy-flavour jets to account for this. The ΔR_{jj} cannot be lower than 0.4 as the jets are reconstructed with a radius of 0.4.

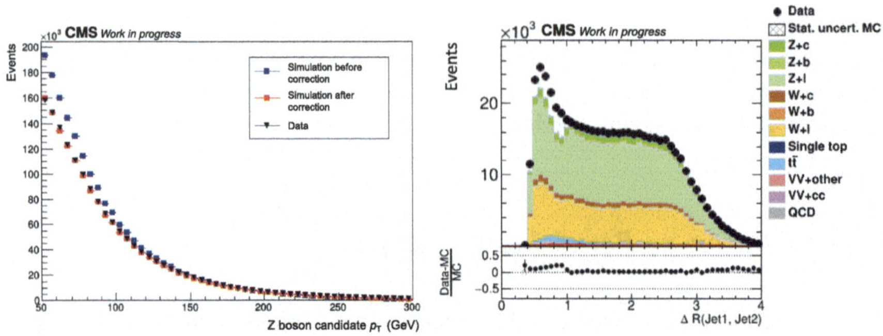

Fig. 9.10 Illustrations of Z boson p_T reweighting in Z+jets simulated samples in the 2L channel (left) and disagreement between data and simulation at low angular separation of two jets in the 0L channel (right)

9.7.6 Electroweak Corrections in VH Signal

The simulated signal sample for the qqVH process is produced with POWHEG+ MiNLO and then rescaled to the production cross section calculated at NNLO in QCD. The total cross section for WH and ZH productions are given by

$$\sigma^{\mathrm{WH}} = \sigma^{\mathrm{qqWH}}_{\mathrm{NNLOinQCD}}(1 + \delta_{\mathrm{EW}}) + \sigma_{\text{t-loop}} + \sigma_{\gamma}, \quad (9.4)$$

$$\sigma^{\mathrm{ZH}} = \sigma^{\mathrm{qqZH}}_{\mathrm{NNLOinQCD}}(1 + \delta_{\mathrm{EW}}) + \sigma_{\text{t-loop}} + \sigma_{\gamma} + \sigma^{\mathrm{ggZH}}, \quad (9.5)$$

where δ_{EW} is an electroweak correction and can be applied as a function of the vector boson p_{T}. Their values have been derived by the LHC Higgs Cross Section Working Group [7] and used in the VHcc analysis.

9.8 Uncertainty Estimation

Several sources of uncertainties (nuisance parameters) have been considered in the analyses. A brief description of each source is provided as follows.

- **Jet energy scale**: Uncertainties from the jet energy scale (JES) [18, 19] are assessed by varying up and down the energy scale of each jet within one standard deviation, individually for each of the 11 sources of uncertainties assessed by the JETMET group. Effects of the variation are propagated to MET as well. These uncertainties are applied to the selected jets after c jet energy regression. Some of these 11 sources are considered correlated across years and others are considered decorrelated, following official prescription of the JETMET POG.

- **Jet energy resolution**: Uncertainties from the jet energy resolution [18, 19] are assessed by varying the resolution scale factors within their uncertainties in the smearing of the reconstructed jets in simulation. A single source of the uncertainty is considered, including both the statistical and systematic uncertainties of the measured resolution scale factors (cf. Sect. 9.7.3). Effects of the variation are also propagated to MET. The uncertainty is applied to selected jets after regression, only in the VH and diboson samples. It is treated as uncorrelated between years.
- **MET unclustered energy**: Uncertainty on the MET related to unclustered energy [19] is assessed by varying each particle type by its own resolution as recommended by the JETMET group. This uncertainty is treated as uncorrelated between years.
- **HEM15/16 issue**: The data collected in 2018 is partially affected by the failure of HEM15/16. In addition to the mitigation measures described in Sect. 9.1.1, an additional uncertainty on the JES is assigned for datasets simulated using 2018 conditions by scaling down the jet energy by 20% for jets with $-1.57 < \phi < -0.87$ and $-2.5 < \eta < -1.3$, and 35% for jets with $-1.57 < \phi < -0.87$ and $-3.0 < \eta < -2.5$, and also propagating it to the MET.
- **Lepton efficiency**: Corrections are applied to the efficiency of the muon and electron trigger, reconstruction, identification, and isolation requirements using data-to-simulation SFs [20–22]. Uncertainties on the measured SFs are evaluated from the statistical uncertainties of the measurement and efficiency differences using alternative samples and selections, and are propagated to the analysis. This uncertainty is treated as uncorrelated between years.
- **MET trigger efficiency**: The efficiency of the MET triggers affects the expected yields of the signals and background processes that are predicted directly from simulated samples in the 0L channel. An uncertainty of 2%, as estimated in the VHbb analysis [15], is applied in the 0L channel. This uncertainty is treated as uncorrelated between years.
- **Pileup reweighting**: Effects due to the uncertainty in the distribution of the number of pileup interactions are evaluated by varying the cross section used to predict the number of pileup interactions in simulation by 4.6% from its nominal value. This uncertainty is treated as uncorrelated between years.
- **Pileup jet identification efficiency**: The efficiencies and misidentification rates of the pileup jet identification algorithm [23] applied to the AK4 jets are corrected by data-to-simulation SFs provided by the JETMET POG. Uncertainties on these SFs are propagated in this analysis. This uncertainty is treated as uncorrelated between years.
- **V p_T reweighting**: NLO electroweak correction parametrised in p_T of V are applied on the V+jets samples. Uncertainty on this NLO electroweak correction is estimated by comparing the shapes with and without applying the correction, for W+jets and Z+jets processes separately. The overall yields are normalized after the baseline selection to factor out the moderate effect on the normalization, as the normalization of the V+jets processes is determined from the fit. This uncertainty is treated as uncorrelated for Z+jets and W+jets processes, but correlated across years.

- **Parton distribution function (PDF)**: Effects of the uncertainty on the PDF is assessed by reweighting each event according to the 30 (100 depending on the sample) replicas of the NNPDF set, and taking the standard deviation of the 30 (100) variations, following the prescription in Ref. [24]. The uncertainty is treated as uncorrelated between different processes, but correlated across years. Depending on the process the uncertainty varies between 1 and 3%.
- **Renormalisation and factorisation scales**: Uncertainties due to the renormalisation scale μ_R and the factorisation scale μ_F are assessed by varying the scales independently by a factor of 0.5 or 2. The effects of such variations are implemented as weights in the event generators. Only effects on the shapes of the final fit variables are considered, i.e. the weights are rescaled to preserve the cross section of each process. The uncertainty is treated as uncorrelated between different processes, but correlated across years.
- **Size of simulated samples**: Uncertainties due to limited size of the simulated samples are taken into account via the Barlow-Beeston method [25]. These are treated as bin-by-bin uncertainties in the fit.
- **Statistical uncertainty of control regions in data**: The limited event yield in the CRs results in an uncertainty on the normalization of the V+jets and $\mathrm{t\bar{t}}$ process measured from the CRs. The uncertainty is taken into account by suitably propagating the uncertainties on the corresponding rate parameters.
- **Cross section of the diboson and SingleTop processes**: The diboson backgrounds are predicted directly from simulation. An uncertainty of 15% is assigned for the diboson processes based on the CMS cross section measurements [26–28]. Similarly an uncertainty of 15% is assigned for the SingleTop processes as well.
- **Luminosity**: A 2.5%, 2.3% and 2.5% uncertainty is assigned on the integrated luminosity measured by CMS for the 2016, 2017, and 2018 data-taking periods [2–4], respectively.
- **Charm tagger reshaping**: The uncertainties derived on the c tagger SFs in Sect. 7.5 have been propagated to the analyses as uncertainties in event weights, for each systematic source. These contribute to several additional sources of uncertainties, each of which is correlated across years. Only the statistical uncertainty in the SFs is considered uncorrelated across years.
- **p_T spectra of DY process**: As described in Sect. 9.7.4, V p_T dependent corrections were derived for DY+jets events using a linear fit in the LF CR. An uncertainty in the correction factor is obtained by varying the fit parameters up/down by their relative uncertainties and replacing the nominal correction factor with its corresponding up/down value. The uncertainty is considered uncorrelated between years.
- **ΔR_jj reweighting of V+jets processes**: As described in Sect. 9.7.5, ΔR_{jj}-dependent corrections are applied to V+jets events for events with $\Delta R_{jj} < 1.0$. An uncertainty on the fit used to derive the correction factors is obtained by varying the fit parameters up/down by their uncertainties. The nominal ΔR_{jj} correction factor is then replaced with a value calculated based on the up/down fit parameters. The uncertainty is uncorrelated between years and between simulated processes (including processes split by flavour compositions).

- **Flavour dependence of ΔR_{jj} reweighting**: A small flavour dependence of the ΔR_{jj} corrections was observed at very low values of ΔR_{jj} (< 0.55). To account for this, flat uncertainties corresponding to twice the maximum relative differences in flavour yields are associated to heavy flavour V+jet components in regions $\Delta R_{jj} < 0.55$. These uncertainties are uncorrelated between years and between simulated processes (including processes split by flavour compositions). The uncertainties used are:
  - Z+c: 20%
  - Z+b: 35%
  - W+c: 30%
  - W+b: 30%.
- **Theoretical uncertainties on the VH cross sections**: Uncertainties on the VH production cross section due to uncertainty of PDF and the choice of the renormalisation and factorization scales are evaluated in Ref. [7] and are taken into account in the VHcc analysis.
- **p_T spectra of VH processes**: The simulated VH samples used in the VHcc analysis are generated at NLO QCD precision. Higher order NLO electroweak and NNLO QCD corrections, parametrised in V p_T, are used to reweight the simulated samples [7]. Uncertainties due to this reweighting are taken into account as a shape uncertainty.
- **Theoretical uncertainties on the** $H \rightarrow c\bar{c}$ **branching ratio**: An uncertainty of +5.5% and –2.0% is used on signal normalization [7]. This uncertainty is treated as uncorrelated between years.

References

1. CMS Collaboration (2016) CMS luminosity measurement for the 2015 data taking period. CMS Physics Analysis Summary. CMS-PAS-LUM-15-001. https://cds.cern.ch/record/2138682
2. CMS Collaboration (2017) CMS luminosity measurements for the 2016 data taking period. CMS Physics Analysis Summary. CMS-PAS-LUM-17-001. https://cds.cern.ch/record/2257069
3. CMS Collaboration (2018) CMS luminosity measurement for the 2017 data-taking period at $\sqrt{s} = 13$ TeV. CMS Physics Analysis Summary. CMS-PAS-LUM-17-004. https://cdsweb.cern.ch/record/2621960
4. CMS Collaboration (2019) CMS luminosity measurement for the 2018 data-taking period at $\sqrt{s} = 13$ TeV. CMS Physics Analysis Summary. CMS-PAS-LUM-18-002. https://cds.cern.ch/record/2676164
5. Hamilton K, Nason P, Zanderighi G (2012) MINLO: multi-scale improved NLO. JHEP 10:155. https://doi.org/10.1007/JHEP10(2012)155. arXiv:1206.3572
6. Luisoni G, Nason P, Oleari C, Tramontano F (2013) $HW^{\pm}$/HZ + 0 and 1 jet at NLO with the POWHEG BOX interfaced to GoSam and their merging within MiNLO. JHEP 10:083. https://doi.org/10.1007/JHEP10(2013)083. arXiv:1306.2542
7. CERN (2017) CERN yellow reports: monographs, vol 2 (2017). Handbook of LHC Higgs cross sections: 4. deciphering the nature of the Higgs sector. https://doi.org/10.23731/CYRM-2017-002, https://e-publishing.cern.ch/index.php/CYRM/issue/view/32

8. Campbell JM, Ellis RK (2010) MCFM for the Tevatron and the LHC. Nucl Phys Proc Suppl 205–206:10. https://doi.org/10.1016/j.nuclphysbps.2010.08.011. arXiv:1007.3492
9. Gehrmann T et al (2014) W^+W^- production at hadron colliders in next to next to leading order QCD. Phys Rev Lett 113:212001. https://doi.org/10.1103/PhysRevLett.113.212001. arXiv:1408.5243
10. Li Y, Petriello F (2012) Combining QCD and electroweak corrections to dilepton production in FEWZ. Phys Rev D 86:094034. https://doi.org/10.1103/PhysRevD.86.094034. arXiv:1208.5967
11. Frixione S et al (2008) Single-top hadroproduction in association with a W boson. JHEP 07:029. https://doi.org/10.1088/1126-6708/2008/07/029. arXiv:0805.3067
12. Re E (2011) Single-top Wt-channel production matched with parton showers using the POWHEG method. Eur Phys J C 71:1547. https://doi.org/10.1140/epjc/s10052-011-1547-z. arXiv:1009.2450
13. Kidonakis N (2014) NNLL threshold resummation for top-pair and single-top production. Phys Part Nucl 45:714. https://doi.org/10.1134/S1063779614040091. arXiv:1210.7813
14. CMS Collaboration (2020) A deep neural network for simultaneous estimation of b jet energy and resolution. Comput Softw Big Sci 4:10. https://doi.org/10.1007/s41781-020-00041-z. arXiv:1912.06046
15. CMS Collaboration (2018) Observation of Higgs boson decay to bottom quarks. Phys Rev Lett 121:121801. https://doi.org/10.1103/PhysRevLett.121.121801. arXiv:1808.08242
16. Rocamora Perez G (2020) Use of Deep Learning techniques and Kinematic Fit with constraints in the search for the $H \to c\bar{c}$ channel at CMS. Presented 27 Nov 2020. https://cds.cern.ch/record/2747746
17. CMS Collaboration (2022) A new calibration method for charm jet identification validated with proton-proton collision events at $\sqrt{s}$ =13 TeV. JINST 17:P03014. https://doi.org/10.1088/1748-0221/17/03/P03014. arXiv:2111.03027
18. CMS Collaboration (2018) Jet energy scale and resolution performance with 13 TeV data collected by CMS in 2016. CMS Detector Performance Note. CMS-DP-2018-028. https://cds.cern.ch/record/2622157
19. CMS Collaboration (2016) Performance of missing energy reconstruction in 13 TeV pp collision data using the CMS detector. CMS Physics Analysis Summary. CMS-PAS-JME-16-004. https://cds.cern.ch/record/2205284
20. CMS Collaboration (2012) Performance of CMS muon reconstruction in pp collision events at $\sqrt{s} = 7$ TeV. JINST 7:P10002. https://doi.org/10.1088/1748-0221/7/10/P10002. arXiv:1206.4071
21. CMS Collaboration (2021) Electron and photon reconstruction and identification with the CMS experiment at the CERN LHC. JINST 16:P05014. https://doi.org/10.1088/1748-0221/16/05/p05014. arXiv:2012.06888
22. CMS Collaboration (2020) Performance of the reconstruction and identification of high-momentum muons in proton-proton collisions at $\sqrt{s} = 13$ TeV. JINST 15:P02027. https://doi.org/10.1088/1748-0221/15/02/P02027. arXiv:1912.03516
23. CMS Collaboration (2020) Performance of the pile up jet identification in CMS for Run 2. CMS Detector Performance Note. CMS-DP-2020-020. https://cds.cern.ch/record/2715906
24. Butterworth J et al (2016) PDF4LHC recommendations for LHC Run II. J Phys G 43:023001. https://doi.org/10.1088/0954-3899/43/2/023001. arXiv:1510.03865
25. Barlow RJ, Beeston C (1993) Fitting using finite Monte Carlo samples. Comput. Phys. Commun. 77:219–228. https://doi.org/10.1016/0010-4655(93)90005-W
26. CMS Collaboration (2016) Measurement of the W^+W^- cross section pp collisions at $\sqrt{s} = 13$ TeV. CMS Physics Analysis Summary. CMS-PAS-SMP-16-006. https://cds.cern.ch/record/2160868
27. CMS Collaboration (2018) Measurements of the $pp \to ZZ$ production cross section and the $Z \to 4\ell$ branching fraction, and constraints on anomalous triple gauge couplings at $\sqrt{s} = 13$ TeV. Eur Phys J C 78:165. https://doi.org/10.1140/epjc/s10052-018-5567-9. arXiv:1709.08601. [Erratum: Eur Phys J C 78:515 (2018)]

28. CMS Collaboration (2022) Measurement of the inclusive and differential WZ production cross sections, polarization angles, and triple gauge couplings in pp collisions at $\sqrt{s} = 13$ TeV. JHEP 07:032. https://doi.org/10.1007/JHEP07(2022)032. arXiv:2110.11231

Chapter 10
Signal Extraction

The signals of the VHcc, VZcc, and VWcq analyses are extracted using separate binned maximum likelihood fits (cf. Sect. 4.7.2). The variable that is used in each fit is one that has different distributions in the signal and in the background. For the VHcc and VZcc analyses, a multivariate analysis technique is implemented by training BDTs (cf. Sect. 4.4,) to separate the respective signals from the backgrounds. The output of the BDT is used as the fit variable in the SRs. On the other hand, the invariant mass of the dijet system (m_{jj}) is used as the fit variable in the VWcq analysis. The former approach usually yields a stronger discrimination against the background and is used when the signal strength is expected to be weak, to obtain more statistically significant results. However, employing a BDT also requires careful scrutiny and calibrations of the input variables. The latter approach provides more interpretable results visualised in terms of intuitive physical observables, at the cost of overall sensitivity.

In the CRs of all the analyses, the calibrated c tagger output (CvsL or CvsB, depending on the region) is used as the fit variable. These variables also provide a separation (in terms of the variable distribution templates) between the background being constrained in each CR and the other major backgrounds in that CR. A good separation allows for a more accurate determination of the rate parameters for the respective backgrounds.

10.1 BDT Training and Performance

Separate BDTs are trained for the VHcc and VZcc analyses, and for different channels and different data-taking years. In the VZcc analysis, only one BDT is trained for the 2LH and 2LL SRs using an inclusive selection criterion, due to the lack of enough simulated signal events. Therefore, the VHcc analysis requires the training of 12 BDTs, while the VZcc requires 9. 50% of the available simulated signal and

S. Mondal, *Charming Decays of the Higgs, Z, and W Bosons*, Springer Theses,
https://doi.org/10.1007/978-3-031-84677-9_10

background events are used in the training, with the exception of V+jets in the simulations of the dataset corresponding to the 2017 data-taking conditions, where approximately 35% is used. Only the remaining simulated events are used in the construction of the SRs used in the fit, to avoid any overlap with the training dataset.

The list of input variables used in each BDT depends on the channel, as listed in Table 10.1. The same variables are used in the same channel across different years of data-taking. The simulated events that pass the selection criteria (cf. Sect. 9.6) of each of the SRs are used in the trainings of the BDTs for the respective SRs, with the exception of 2LL and 2LH being merged in the VZcc analysis. Each event is weighted by the cross section of the process (cf. Eq. 4.1) and the correction weights (cf. Sect. 9.7) so that the relative yields of the different background processes are representative of the SR contributions in the training samples.

The performance of the trained BDTs can be examined by plotting the ROC curves of the discriminators. The distributions of the signal and background samples as a function of the BDT output, as well as the corresponding ROC curves for the four (three) BDTs used in the VHcc (VZcc) analysis are shown in Figs. 10.1 and 10.2. These figures show that the distributions of the BDT output scores are very different for the signal and the background events in all channels.

10.2 VHcc Signal Extraction

This section summarises the results of the binned maximum likelihood fit with the CL_S method (cf. Sect. 4.7.2) using the VHcc process as the signal. The signal extraction is performed with a simultaneous fit of the CRs and the SRs in 4 channels, using data from the three data-taking years. The BDT outputs are used as the fit variable in the SRs. The CvsB discriminator is used in the HF and CC CRs and a single bin is used in the TT and LF CRs. This is because while the TT and LF CRs are dominated respectively by $t\bar{t}$ and V+light jet processes, the HF and CC CRs have significant contaminations from processes other than V+b and V+c jets and therefore the different templates for b-enriched and c-enriched processes provided by the CvsB discriminator in each CR help in constraining the rate parameters of the processes more efficiently.

The cross section of the VHbb process in the fit is set to the value expected from the SM. It has been verified that varying the cross section of the VHbb process changes the expected sensitivity of the VHcc measurement by $\sim$5%.

The upper limit on the signal strength μ for the production of the VHcc process is extracted at 95% CL based on the CL_S method under the asymptotic approximation for the profile likelihood test statistic. The same measurement is repeated with a p_T requirement of < 300 GeV on an AK15 jet (p_T^{AK15}) in the event, if reconstructed. The second approach enables combining the results with the complementary merged-jet approach. The results are discussed separately in the following paragraphs.

Table 10.1 Variables used as inputs to the BDT used for each channel of the VHcc and VZcc analyses for 2016, 2017, and 2018 data-taking periods. The 2LL and 2LH channels in the VHcc analysis have separate BDT trainings, but the BDTs have the same input variables, as listed in the 2L column

Variable	Description	Used in...?		
		0L	1L	2L
m_{jj}	Dijet invariant mass	✓	✓	✓
$p_{T,jj}$	Dijet p_T		✓	✓
V p_T	Vector boson p_T		✓	✓
V m_T	Vector boson transverse mass		✓	
MET	Missing transverse energy	✓	✓	
V $p_T/p_{T,jj}$	Ratio of V p_T to $p_{T,jj}$	✓	✓	✓
Jet 1 CvsL	CvsL of leading jet	✓	✓	✓
Jet 1 CvsB	CvsB of leading jet	✓	✓	✓
Jet 2 CvsL	CvsL of subleading jet	✓	✓	✓
Jet 2 CvsB	CvsB of subleading jet	✓	✓	✓
Jet 1 p_T	p_T of leading jet	✓	✓	✓
Jet 2 p_T	p_T of subleading jet	✓	✓	✓
$\Delta\phi_{V,jj}$	Azimuthal angle between V and dijet	✓	✓	✓
$\Delta R_{j_1,j_2}$	ΔR between leading and subleading jets		✓	✓
$\Delta\phi_{j_1,j_2}$	$\Delta\phi$ between leading and subleading jets	✓	✓	
$\Delta\eta_{j_1,j_2}$	$\Delta\eta$ between leading and subleading jets	✓	✓	✓
$\Delta\phi_{l_1,l_2}$	$\Delta\phi$ between selected prompt leptons			✓
$\Delta\eta_{l_1,l_2}$	$\Delta\eta$ between selected prompt leptons			✓
$\Delta\phi_{l_1,j_1}$	$\Delta\phi$ between higher-p_T lepton and leading jet		✓	
$\Delta\phi_{l_2,j_1}$	$\Delta\phi$ between lower-p_T lepton and leading jet			✓
$\Delta\phi_{l_2,j_2}$	$\Delta\phi$ between lower-p_T lepton and subleading jet			✓
$\Delta\phi_{l_1,\mathrm{MET}}$	$\Delta\phi$ between higher-p_T lepton and MET		✓	
$\Delta\eta_{l_1,\mathrm{top}}$	$\Delta\eta$ between higher-p_T lepton and top candidate		✓	
$\Delta\phi_{l_1,\mathrm{top}}$	$\Delta\phi$ between higher-p_T lepton and top candidate		✓	
$\Delta R_{l_1,\mathrm{top}}$	ΔR between higher-p_T lepton and top candidate		✓	
$\mathrm{CvsL_{top}}$	CvsL score of b jet candidate in the top reco		✓	

(continued)

Table 10.1 (continued)

Variable	Description	Used in...?		
		0L	1L	2L
$\mathrm{CvsB}_{\mathrm{top}}$	CvsB score of b jet candidate in the top reco		✓	
$\mathrm{B}_{\mathrm{top}}$	BvsC/L score of b jet candidate in the top reco		✓	
m_{top}	Mass of the reconstructed top candidate		✓	
N_{aj}	Number of additional jets in the event		✓	
$\sigma_{j_1,\mathrm{cReg}}$	j_1 energy resolution from c jet energy regression	✓	✓	✓
$\sigma_{j_2,\mathrm{cReg}}$	j_2 energy resolution from c jet energy regression	✓	✓	✓
$m_{jj}\|_{\mathrm{kinfit}}$	m_{jj} after kinematic fit			✓
$p_{\mathrm{T},jj}\|_{\mathrm{kinfit}}$	$p_{\mathrm{T},jj}$ after kinematic fit			✓
$\Delta\phi_{\mathrm{V},jj}\|_{\mathrm{kinfit}}$	$\Delta\phi_{\mathrm{V},jj}$ after kinematic fit			✓
$\Delta\eta_{\mathrm{V},jj}\|_{\mathrm{kinfit}}$	$\Delta\eta$ between V and dijet after kinematic fit			✓
Jet 1 $p_{\mathrm{T}}\|_{\mathrm{kinfit}}$	Leading jet p_{T} after kinematic fit			✓
Jet 2 $p_{\mathrm{T}}\|_{\mathrm{kinfit}}$	Subleading jet p_{T} after kinematic fit			✓
$\sigma_{m_{jj}}\|_{\mathrm{kinfit}}$	Resolution in m_{jj} after kinematic fit			✓

- **p_{T}-inclusive VHcc analysis**: The rate parameters of each background constrained by the fit are listed in Table 10.2. The expected ("blinded"; cf. Sect. 4.7.2) upper limits on the signal strength modifier for the p_{T}-inclusive topology per channel and year of data-taking along with the combined upper limit are presented in Table 10.3. The combined upper limit using the full Run-2 data is 13.9 (expected) at the 95% CL. These results are not unblinded as the final strategy of the analysis is to combine this topology with the merged-jet approach. However, these numbers give an estimate of the sensitivity of the analysis using the resolved-jet topology alone.
- **VHcc analysis with $p_{\mathrm{T}}^{\mathrm{AK15}} < 300$ GeV**: The expected and observed upper limits on the signal strength modifier for the $p_{\mathrm{T}}^{\mathrm{AK15}} < 300\,\mathrm{GeV}$ topology per channel and year of data-taking along with the combined upper limits are presented in Table 10.4. The observed (expected) combined upper limit using the full Run-2 data is 14 (19) at the 95% CL. The BDT distribution in some of the SRs are shown in Fig. 10.3. These plots show a good agreement between post-fit simulation and data. The simulated background event yield falls sharply with increasing BDT score, thereby making the event yields corresponding to high BDT scores sensitive to the signal strength.

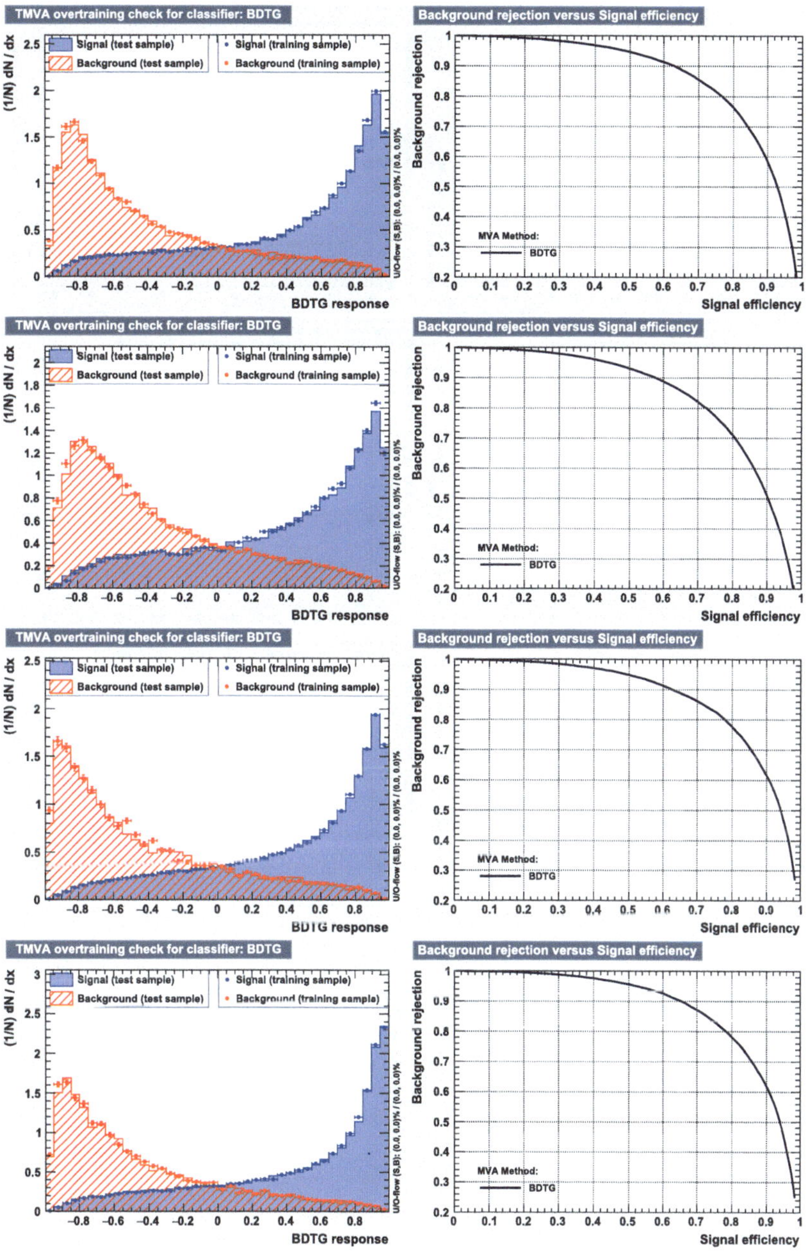

Fig. 10.1 Distributions of VHcc signal (blue) and background (red, hatched) events in the BDT output (right) and the corresponding ROC curves (right) for the VHcc analysis with samples simulated using the 2018 data-taking conditions, in the 0L (first row), 1L (second row), 2LL (third row), and 2LH (fourth row) channels. BDTG stands for "BDT (Gradient-boosted)". The default range of BDT outputs in TMVA is from −1 to +1, but has been transformed to 0−1 in the analysis using BDT = $\frac{\text{BDTG}+1}{2}$

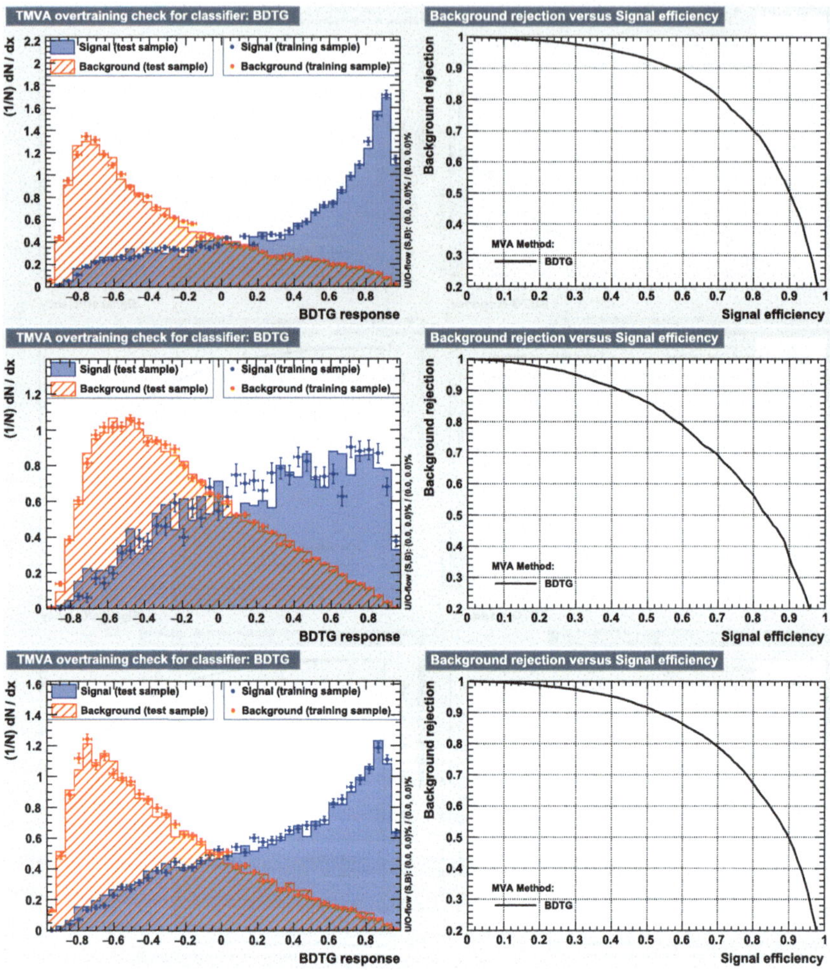

Fig. 10.2 Distributions of VZcc signal (blue) and background (red, hatched) events in the BDT output (right) and the corresponding ROC curves (right) for the VZcc analysis with samples simulated using the 2018 data-taking conditions, in the 0L (first row), 1L (second row), and 2L (third row) channels. BDTG stands for "BDT (Gradient-boosted)". The default range of BDT outputs in TMVA is from -1 to $+1$, but has been transformed to 0–1 in the analysis using BDT $= \frac{\text{BDTG}+1}{2}$

Table 10.2 Background normalization rate parameters determined from the fit across all data-taking years and all channels of VHcc analysis, inclusively in AK15 p_T. The same RPs were used for W+jet events in 0L and 1L channels

Process	2016	2017	2018	Applicable to...			
				0L	1L	2LL	2LH
Z($\nu\nu$)+c	0.92 ± 0.20	1.11 ± 0.12	1.10 ± 0.21	✓			
Z($\nu\nu$)+b	0.87 ± 0.18	1.12 ± 0.21	0.80 ± 0.21	✓			
Z($\nu\nu$)+light	1.10 ± 0.14	1.67 ± 0.11	0.70 ± 0.12	✓			
W+c	0.80 ± 0.10	0.98 ± 0.22	1.17 ± 0.17	✓	✓		
W+b				✓	✓		
W+light	1.05 ± 0.07	1.05 ± 0.12	0.84 ± 0.10	✓	✓		
Z(ll)+c (2LL)	0.84 ± 0.11	0.77 ± 0.09	0.80 ± 0.26			✓	
Z(ll)+b (2LL)	0.80 ± 0.05	0.89 ± 0.06	0.97 ± 0.08			✓	
Z(ll)+light (2LL)	1.00 ± 0.06	0.95 ± 0.06	0.96 ± 0.09			✓	
Z(ll)+c (2LH)	0.70 ± 0.10	0.70 ± 0.11	0.76 ± 0.13				✓
Z(ll)+b (2LH)	0.89 ± 0.07	1.01 ± 0.12	1.08 ± 0.12				✓
Z(ll)+light (2LH)	1.02 ± 0.06	0.83 ± 0.08	0.88 ± 0.09				✓
$t\bar{t}$ (0L)	0.79 ± 0.07	0.96 ± 0.07	0.88 ± 0.08	✓			
$t\bar{t}$ (1L)	0.85 ± 0.06	1.00 ± 0.06	0.98 ± 0.07		✓		
tt (2LL)	0.85 ± 0.05	0.86 ± 0.05	0.88 ± 0.05			✓	
$t\bar{t}$ (2LH)	0.80 ± 0.07	0.85 ± 0.06	0.88 ± 0.07				✓

Table 10.3 Expected upper limits at the 95% CL on the signal strength modifier μ for the VHcc process with the p_T-inclusive topology

Year	0L	1L	2L	Combined
2016	45.0	62.0	33.4	24.6
2017	46.5	53.4	35.1	24.7
2018	43.9	42.1	34.1	22.6
Full Run-2	26.9	30.3	19.1	13.9

Table 10.4 Expected and observed upper limits at the 95% CL on the signal strength modifier μ for the VHcc process with the $p_{\mathrm{T}}^{\mathrm{AK15}} < 300\,\mathrm{GeV}$ topology

Year	Type	0L	1L	2L	Combined
2016	Expected	72	87	41	34
	Observed	58	63	46	28
2017	Expected	63	73	40	33
	Observed	39	71	40	24
2018	Expected	65	67	43	33
	Observed	56	50	46	27
Full Run-2	Expected	38	43	23	19
	Observed	25	33	23	14

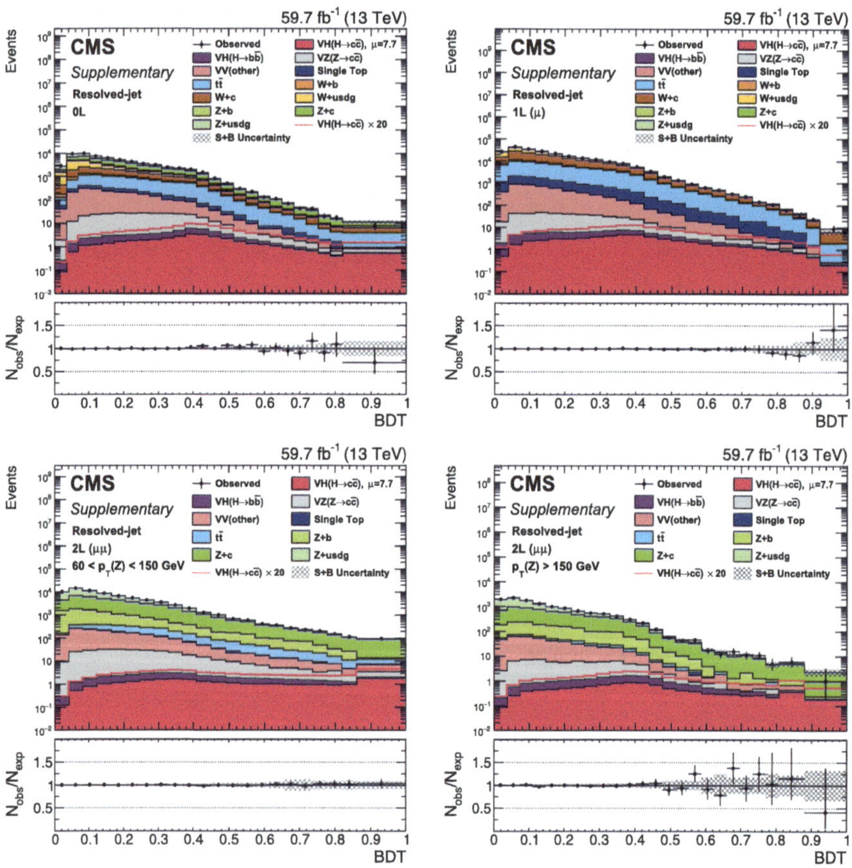

Fig. 10.3 Post-fit BDT distributions in the SRs of the 0L (upper-left), 1L(μ) (upper-right), 2LL($\mu\mu$) (lower-left), and 2LH($\mu\mu$) (lower-right) channels of the VHcc analysis in the $p_{\mathrm{T}}^{\mathrm{AK15}} < 300\,\mathrm{GeV}$ topology with the dataset collected in 2018

The rate parameters of the background processes in the $p_T^{AK15} < 300$ GeV case were observed to be close to those of the p_T-inclusive case, well-compatible within uncertainties.

10.3 VZcc Signal Extraction

The VZcc signal extraction is also performed with a binned maximum likelihood fit. However, instead of using the CL_S method, the significance of the signal is evaluated in this case. Just like in the case of the VHcc analysis, the significances in the VZcc analysis are evaluated separately for the p_T-inclusive topology and for the $p_T^{AK15} < 300$ GeV case. The results are reported in the following paragraphs.

- **p_T-inclusive VZcc analysis**: The expected (blinded) significances of the VZcc signal over the background-only prediction for the p_T-inclusive topology per channel and year of data-taking along with the combined significance are presented in Table 10.5. The combined significance using the full Run-2 data is 3.8σ.
- **VZcc analysis with $p_T^{AK15} < 300$ GeV**: The expected and observed significances of the VZcc signal over the background-only prediction for the $p_T^{AK15} < 300$ GeV topology per channel and year of data-taking along with the combined significances are presented in Table 10.6. The observed (expected) combined significance using the full Run-2 data is 3.1σ (3.3σ) over the background-only prediction.

Table 10.5 Expected significances in units of σ of the VZcc process over the background-only prediction with the p_T-inclusive topology

Year	0L	1L	2L	Combined
2016	0.8	0.6	1.6	1.6
2017	1.5	0.9	1.8	2.6
2018	1.4	0.9	1.3	2.3
Full Run-2	2.2	1.3	2.7	3.8

Table 10.6 Expected and observed significances in units of σ of the VZcc process over the background-only prediction with the $p_T^{AK15} < 300$ GeV topology

Year	Type	0L	1L	2L	Combined
2016	Expected	0.7	0.5	1.6	1.7
	Observed	2.0	0.0	1.0	1.8
2017	Expected	1.2	0.7	1.8	2.2
	Observed	0.6	0.2	1.5	1.7
2018	Expected	0.8	0.6	1.3	1.7
	Observed	0.0	1.6	0.6	1.9
Full Run-2	Expected	1.6	1.1	2.7	3.3
	Observed	1.3	1.4	1.8	3.1

The rate parameters of the background processes in the VZcc analysis were identical to those in the VHcc ones in most cases, and well-compatible within uncertainties.

10.4 Merged-Jet Analysis and Combination

10.4.1 Merged-Jet Results

The merged-jet analysis strategy has been briefly discussed in Sect. 8.3. However, this thesis does not discuss the details of the methodologies adopted in it. More details can be found in Refs. [1, 2]. This subsection discusses the fit results obtained in the merged-jet VHcc and VZcc analyses.

The expected and observed upper limits on the VHcc signal strength modifier in the merged-jet topology with a requirement of $p_T^{AK15} > 300\,\text{GeV}$, per year of data-taking are presented in Table 10.7. The same table also shows the expected and observed significances of the VZcc signal over the background-only prediction in the merged-jet topology with a requirement of $p_T^{AK15} > 300\,\text{GeV}$, per year of data-taking. The observed (expected) upper limit on the VHcc signal strength modifier using the full Run-2 data is 17 (8.8) at the 95% CL, while the observed (expected) significance of the VZcc process using the full Run-2 data is 4.4σ (4.7σ) over the background-only prediction.

10.4.2 Combination

Results from the resolved-jet and merged-jet topologies are statistically combined to further improve the signal sensitivity. This subsection discusses the combination of the resolved-jet and merged-jet topologies, closely following the results published in

Table 10.7 Expected and observed upper limits to the signal strength modifier μ of the VHcc process at the 95% CL, and the significances of the VZcc process (in units of σ) over the background-only prediction with the merged-jet toplogy with a requirement of $p_T^{AK15} > 300\,\text{GeV}$

Year	Type	VHcc limits	VZcc significances
2016	Expected	19	2.4
	Observed	20	2.0
2017	Expected	17	2.9
	Observed	24	2.6
2018	Expected	14	3.0
	Observed	26	3.1
Full Run-2	Expected	8.8	4.7
	Observed	17	4.4

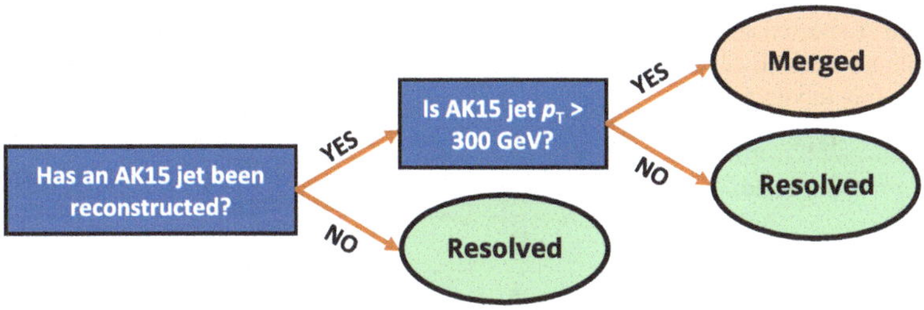

Fig. 10.4 A flowchart showing the categorisation of an event as a candidate for the resolved-jet or merged-jet topology

Ref. [1]. The sets of events selected by the two topologies are kept mutually exclusive by assigning complementary requirements on the p_T of an AK15 jet (p_T^{AK15}), if reconstructed. If multiple AK15 jets are reconstructed in an event, the p_T of the one with the highest p_T is considered for the categorisation. If no AK15 jet is reconstructed in an event, the event is considered as a candidate for the resolved-jet topology. The event categorisation scheme is illustrated in Fig. 10.4. The criterion of 300 GeV for p_T^{AK15} was optimised to maximise the sensitivity of the VHcc and VZcc searches while keeping the fit robust and stable.

The correlation scheme followed for the treatment of the systematic uncertainties between the two topologies is summarized as follows:

- The theoretical uncertainties are treated as fully correlated between the resolved-jet and merged-jet topologies.
- The rate parameters scaling the V+jets and $t\bar{t}$ backgrounds are treated as uncorrelated between the two topologies. This is because the two topologies probe mutually-exclusive phase spaces and have different approaches towards estimating the flavour compositions of the major backgrounds.
- Uncertainties related to the c tagging SFs are treated as uncorrelated, as the procedures followed for the calibration of the two algorithms (DeepJet AK4 and ParticleNet AK15) are significantly different, and also probe very different phase spaces.
- Experimental uncertainties related to JES, JER, MET, pileup, and lepton efficiencies are treated as correlated between the two topologies as they encapsulate the same effects.

Results

The final expected and observed upper limits on the VHcc signal strength modifier at the 95% CL after combining the resolved-jet and merged-jet topologies, along with the best-fit value of the signal strength modifier (μ) per channel and per data-taking year are shown in Table 10.8. The combined observed (expected) upper limit is 14 (7.6) and the corresponding best-fit signal strength is $7.7^{+3.8}_{-3.5}$ ($1.0^{+3.7}_{-3.4}$). The observed

Table 10.8 Expected and observed upper limits on the signal strength modifier μ at the 95% CL along with the best-fit value of μ for the VHcc process after the combination of the resolved-jet and merged-jet topologies

Year	Type	0L	1L	2L	Combined	Best-fit μ
2016	Expected	26	25	30	16	$1.0^{+7.6}_{-7.1}$
	Observed	21	43	21	18	$3.2^{+7.4}_{-7.3}$
2017	Expected	24	25	26	15	$1.0^{+6.9}_{-6.1}$
	Observed	29	26	31	18	$4.3^{+6.9}_{-6.2}$
2018	Expected	20	18	24	12	$1.0^{+5.6}_{-4.9}$
	Observed	34	18	42	21	$9.9^{+6.1}_{-5.1}$
Full Run-2	Expected	13	12	14	7.6	$1.0^{+3.7}_{-3.4}$
	Observed	18	19	20	14	$7.7^{+3.8}_{-3.5}$

upper limit is compatible with the SM expectation of $\mu = 1$ within a 2σ uncertainty. The expected and observed upper limits along with the uncertainties in the expected upper limits are summarised in Fig. 10.5.

The final expected and observed significances of the VZcc process over the background-only expectation after combining the resolved-jet and merged-jet topologies, along with the best-fit values of the signal strength modifier (μ) per channel and per year are shown in Table 10.9. The combined observed (expected) significance is 5.7σ (5.9σ) and the best-fit signal strength is $1.01^{+0.23}_{-0.21}$ ($1.00^{+0.22}_{-0.20}$). This corresponds to an observation of the VZcc process with 5.7σ significance.

Discussion

It is interesting to note from the expected limits/significances in Tables 10.4, 10.6, and 10.7, that the merged-jet topology is more sensitive to the VHcc and VZcc processes compared to the resolved-jet topology in the respective phase spaces. This can be attributed to the fact that the merged-jet approach efficiently exploits correlations between the hadronisation products of the two c quarks, and that the signal reconstruction efficiency is high and background contamination is low at high Lorentz boosts of the Higgs/Z boson candidate. The novel ParticleNet algorithm also helps maximally exploit flavour-related information in jets to reject most light-flavour and bottom jets in the signal regions of the merged-jet topology. However, the observed upper limit on the VHcc signal is more stringent in the resolved-jet topology. This is a result of a downward (upward) statistical fluctuation of $\sim 1\sigma$ ($\sim 2\sigma$) of the observed limit with respect to the expected limit in the resolved-jet (merged-jet) topology of the analysis, as evident in Fig. 10.5.

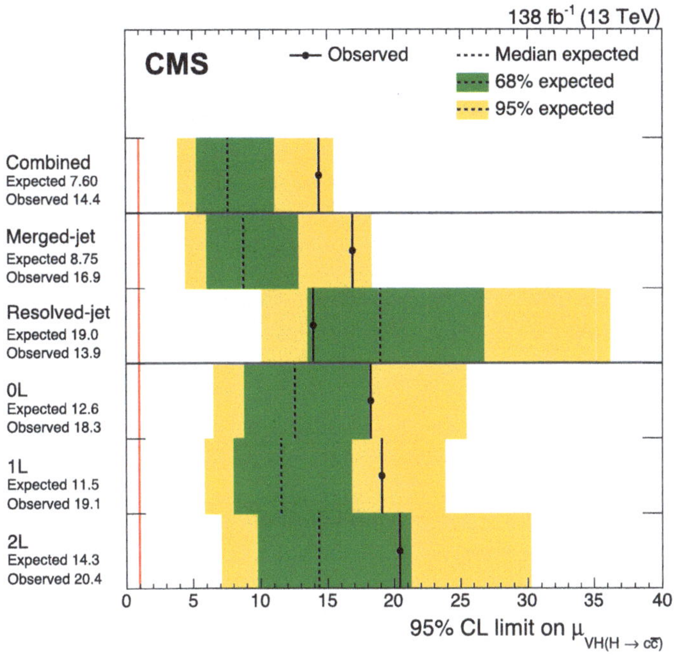

Fig. 10.5 The 95% CL upper limits on the signal strength modifier $\mu_{\mathrm{VH(H}\rightarrow\mathrm{c\bar{c})}}$ of the VHcc process across various channels, the two topologies, and the combined result. Green and yellow bands indicate the expected 68% and 95% CL regions, respectively, under the background-only prediction. The vertical red line indicates the SM value $\mu_{\mathrm{VH(H}\rightarrow\mathrm{c\bar{c})}} = 1$ [1]

Table 10.9 Expected and observed significances (in units of σ) over the background-only prediction, along with the best-fit value of the signal strength modifier (μ), for the VZcc analysis after the combination of the resolved-jet and merged-jet topologies

Year	Type	0L	1L	2L	Combined	Best-fit μ
2016	Expected	2.5	1.0	1.9	3.0	$1.00^{+0.42}_{-0.36}$
	Observed	3.6	1.0	0.4	2.9	$1.02^{+0.41}_{-0.37}$
2017	Expected	2.9	1.3	2.6	3.7	$1.00^{+0.35}_{-0.30}$
	Observed	1.3	0.0	4.0	3.2	$0.84^{+0.32}_{-0.28}$
2018	Expected	3.1	1.4	2.0	3.6	$1.00^{+0.39}_{-0.32}$
	Observed	4.0	1.4	1.2	3.8	$1.23^{+0.45}_{-0.37}$
Full Run-2	Expected	4.8	2.1	3.7	5.9	$1.00^{+0.22}_{-0.20}$
	Observed	4.9	1.3	3.3	5.7	$1.01^{+0.23}_{-0.21}$

These results can be compared with those from ATLAS [3]. The ATLAS VHcc search using full Run-2 data yields an observed (expected) upper limit of 26 (31) times the SM expectation at the 95% CL. Thus the observed (expected) upper limit reported by CMS is about 2 (4) times as strong as the one reported by ATLAS. Similarly, while the ATLAS VZcc search yields an observed (expected) VZcc significance of 2.6σ (2.2σ) above the background-only prediction, the CMS search yields a significance that is about 2 (2.7) times as strong. The difference in sensitivities between the two experiments can be attributed to a number of factors, such as:

- The CMS search splits the topology into two parts (resolved- and merged-jet) and utilises optimised hadronically-decaying Higgs/Z boson reconstruction strategies in the respective toplogies.
- The resolved-jet approach in the CMS analysis utilises a full discriminator calibration approach (cf. Sect. 7), which helps in maximally exploiting the outputs of the charm tagging algorithms compared to traditional calibration techniques implemented in the ATLAS search.
- The resolved-jet approach in the CMS analysis utilises a BDT, that separates signal and background, as the variable used in the signal extraction fit. This helps enhance the separation between the signal and background processes compared to using the dijet invariant mass, as done in the ATLAS search.
- The merged-jet approach in the CMS analysis implements the ParticleNet flavour tagging algorithm (cf. Sect. 8.3), which uses state-of-the-art DGCNN architectures, to identify charm jets more efficiently than algorithms used in the ATLAS search.
- Several novel analysis techniques (cf. Sects. 8.3 and 9.4), namely charm jet energy regression, DGCNN-based mass regression, etc., are implemented in the CMS search to further improve the mass resolution of the reconstructed Higgs/Z boson candidate.
- The ATLAS search also simultaneously combines [3] the results of the VHcc search with the ATLAS VHbb search [4], and, therefore, vetoes parts of the phase space in the VHcc analysis that overlap with that of the VHbb analysis. Such a combination is not performed in the CMS search and hence the CMS search probes a slightly larger part of the signal phase space.

The results of the presented VHcc search can also be compared with the expected sensitivity of such a search at the future High-Luminosity Large Hadron Collider (HL-LHC) [5]. Assuming an integrated luminosity of 3000 fb^{-1} of pp collision data at $\sqrt{s} = 14\,\text{TeV}$ at the HL-LHC, results from the previous ATLAS VHcc searches [3, 6] can be extrapolated to obtain an expected upper limit of around 6 times the SM expectation [7, 8] at the 95% CL at the HL-LHC. A comparable expected upper limit (7.6 times the SM expectation) is obtained with the present search at CMS with 138 fb^{-1} of pp collision data, attesting to the significant improvements in charm tagging capabilities and the successful deployment of modern analysis techniques in the presented analysis.

Combined Plots

As the resolved-jet and merged-jet approaches use different fit variables (BDT and m_{AK15}, respectively), combined SR plots are generated by plotting event yields as a function of S/B, where S and B are the post-fit signal and background yields, respectively, in each bin of the fitted m_{AK15} or BDT discriminant distributions in the respective approach. Figure 10.6 shows the combined results across all channels in all data-taking years for both the VHcc and VZcc analyses.

The plot in Fig. 10.6 corresponding to the VHcc analysis shows a good agreement of the observed data with both the B-only and S+B hypothesis (cf. Sect. 4.7.2), within the uncertainties in the data and simulations. This demonstrates that the analysis is not sensitive enough to discard the B-only hypothesis with the data currently available. The plot corresponding to the VZcc analysis, on the other hand, shows a good agreement of the observed data with the S+B hypothesis, and a stark difference of the data from the B-only simulation that is not covered by the uncertainties. This lets us discard the B-only hypothesis in favour of the S+B hypothesis with a 5.7σ confidence, as measured with statistical methods.

Uncertainties

A breakdown of the uncertainties is presented in Table 10.10 by evaluating the relative impact of each uncertainty on the fitted μ in the VHcc search. The largest contribution is from the statistical uncertainties (~85%) because of the limited number of events in each bin of the fitted SRs (~77%) as well as the limited number of events in the CRs used to extract the background normalisations (~37%). The

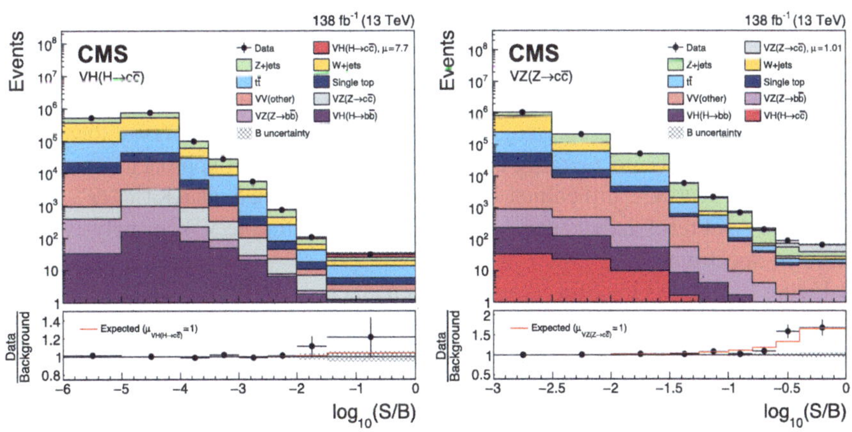

Fig. 10.6 Distribution of events as a function of S/B in the VHcc (left) and VZcc (right) searches, where S and B are the post-fit signal and background yields, respectively, in each bin of the fitted m_{AK15} or BDT distributions after combining the resolved-jet and merged-jet topologies. The red line in the lower panel represents background plus SM signal divided by background

Table 10.10 The relative contributions to the total uncertainty on the signal strength modifier μ for the VHcc process

Uncertainty source		$\frac{\Delta\mu}{\Delta\mu_{\text{total}}}$
Statistical	Background normalisations	37%
	Bin-by-bin uncertainties in the SR	77%
	Total	**85%**
Experimental	Sizes of simulated samples	37%
	Charm identification	23%
	JES and JER	15%
	Simulation modeling	11%
	Luminosity	6%
	Lepton identification	4%
	Total	**48%**
Theory	Backgrounds	17%
	Signal	15%
	Total	**22%**

experimental uncertainties are dominated by the limited number of simulated events (~37%) and charm jet identification uncertainties (~23%). Theoretical uncertainties include the uncertainties in the cross sections, p_T spectra, PDFs, renormalisation and factorisation scales of both the signal and background samples. These represent ~22% of the total uncertainty.

10.5 VWcq Signal Extraction

The signal of the VWcq analysis is extracted using a binned maximum likelihood fit and the significance is reported. Unlike the VHcc and VZcc analyses, the VWcq analysis is not split into p_T ranges, and only the resolved-jet approach is used. As explained in Sect. 8.3, difficulties in calibrating a hypothetical AK15 "cs"-tagger makes the AK4 resolved-jet approach a suitable choice for the VWcq analysis in the entire p_T spectrum.

Figure 10.7 shows the SRs across all the channels of the VWcq analysis with the dataset collected in 2018. The final expected and observed significances of the VWcq process over the background-only expectation, along with the best-fit value of the signal strength modifier (μ) per channel and per year are shown in Table 10.11. The observed (expected) significance is 5.6σ (5.7σ) and the corresponding best-fit signal strength is $1.02^{+0.20}_{-0.19}$ ($1.00^{+0.20}_{-0.18}$). This corresponds to an observation of the VWcq process with 5.6σ significance.

The plots in Fig. 10.7 show a good agreement of the observed data with the S+B hypothesis. While individual plots may not be sufficient to rule out the B-only hypoth-

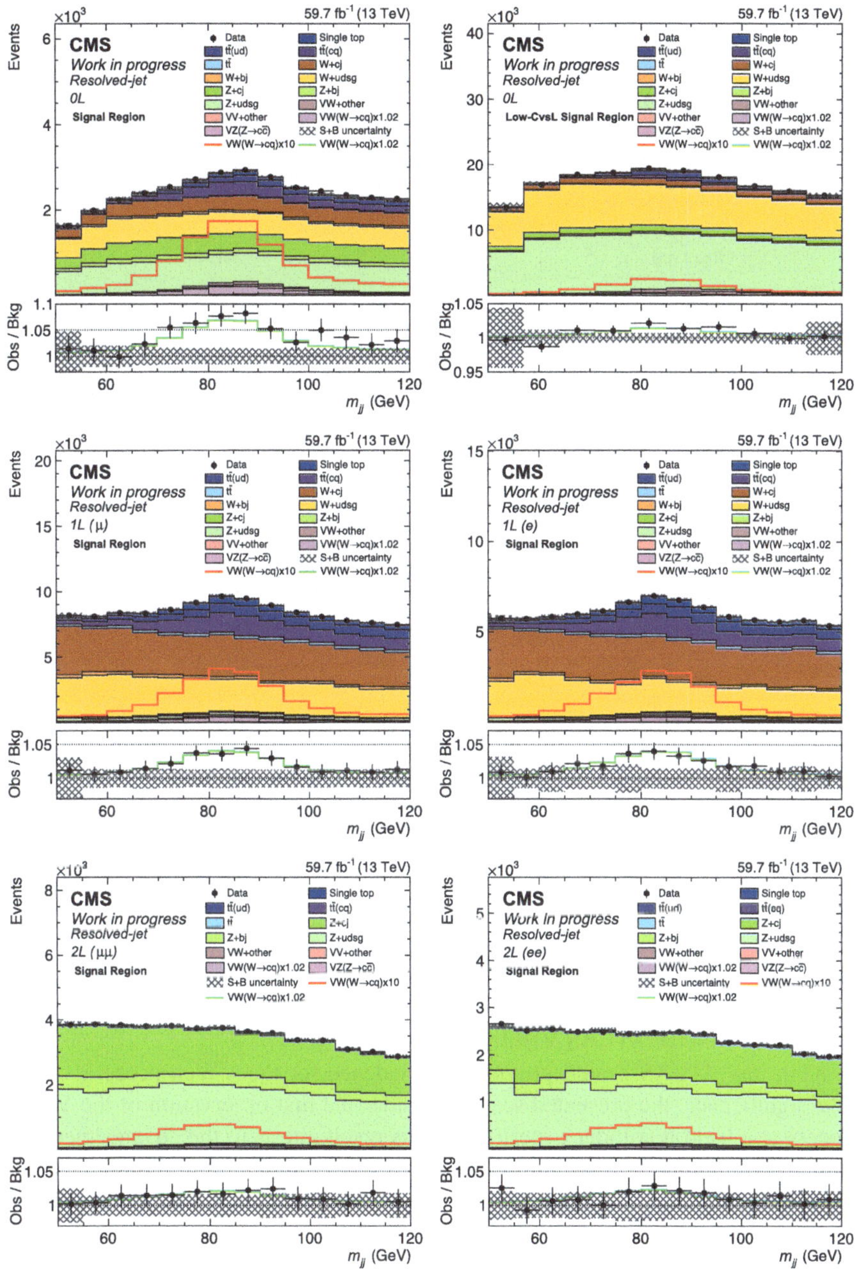

Fig. 10.7 Post-fit m_{jj} distributions in the 0L SR (upper-left), 0L low-CvsL SR (upper-right), 1L(μ) SR (middle-left), 1L(e) SR (middle-right), 2L($\mu\mu$) SR (lower-left), and 2L(ee) SR (lower-right) of the VWcq analysis with the dataset collected in 2018. The red line in the upper panel represents the signal multiplied by 10, while the green line in the lower panel represents post-fit background plus SM signal divided by background

Table 10.11 Expected and observed significances (in units of σ) over the background-only prediction, along with the best-fit value of the signal strength modifier (μ), for the VWcq analysis

Year	Type	0L	1L	2L	Combined	Best-fit μ
2016	Expected	2.0	1.8	0.8	2.9	$1.00^{+0.36}_{-0.35}$
	Observed	1.1	0.2	0.0	1.3	$0.44^{+0.33}_{-0.34}$
2017	Expected	2.3	2.2	0.9	3.3	$1.00^{+0.32}_{-0.30}$
	Observed	2.5	0.8	1.6	3.1	$1.01^{+0.34}_{-0.33}$
2018	Expected	2.8	1.7	0.8	3.6	$1.00^{+0.31}_{-0.29}$
	Observed	5.0	1.9	0.6	5.1	$1.49^{+0.34}_{-0.31}$
Full Run-2	Expected	4.3	3.2	1.5	5.7	$1.00^{+0.20}_{-0.18}$
	Observed	5.4	1.7	1.0	5.6	$1.02^{+0.20}_{-0.19}$

esis with the associated uncertainties, a statistical measurement after combining all the SRs across the three years of data-taking results in a 5.6σ difference of the observed data from the B-only hypothesis, as presented in Table 10.11. Furthermore, the signal is seen to peak at $m_{jj} \approx 80\,\text{GeV}$, corresponding to the mass of the W boson.

This result can be compared with the corresponding result from the ATLAS experiment [3]. The ATLAS search yields an observed (expected) VWcq significance of 3.8σ (4.6σ) above the background-only prediction at the 95% CL, while the presented search with CMS data yields a significance that is about 1.5 (1.2) times as strong. In this case, the hadronically-decaying W boson reconstruction strategy is similar (resolved-jet topology only) and the fit variable in the SRs are the same (m_{jj}). However, the use of full discriminator calibration approach (cf. Sect. 7) and the several improved analysis techniques (cf. Sect. 9.4) used in the presented VWcq analysis help achieve a stronger significance compared to the ATLAS search. Moreover, the ATLAS search explicitly excludes parts of the phase space in the analysis that overlap with the ATLAS VHbb analysis [4].

While the ATLAS search provides an evidence for the VWcq process with a 3.8σ significance, the presented search provides the first observation of the VWcq process at a hadron collider experiment. This result attests to the advancements in charm tagging capabilities and the unprecedented sensitivities achievable at the CMS experiment.

The correlated charm-strange (cs) tagging approach (cf. Sect. 8.2) used in this search may also be extended to other searches in the future, such as those for hypothetical charged Higgs bosons decaying to charm and strange quarks. The same cs tagging strategy can also be used to precisely measure the charm decay fraction of the W boson, or the ratio $R_c^W = \frac{\Gamma(W\to cq)}{\Gamma(W\to \text{hadrons})}$ of the charmed W decay width to the total hadronic decay width of the W boson [9, 10]. Such a measurement at hadron colliders is expected to be more sensitive in other channels, such as $t\bar{t}$, than the low-sensitivity diboson channel that this thesis discusses. Measurement of this ratio also allows for a precise measurement of the SM Cabibbo-Kobayashi-Maskawa (CKM) matrix parameter, $|V_{cs}|$.

References

1. CMS Collaboration Search for Higgs boson decay to a charm quark-antiquark pair in proton-proton collisions at $\sqrt{s}$ = 13 TeV. https://doi.org/10.48550/ARXIV.2205.05550. arXiv:2205.05550. Accepted for publication by PRL
2. Qu H (2019) A search for Lorentz-boosted Higgs bosons decaying to charm quarks in the CMS experiment using deep neural networks. http://cds.cern.ch/record/2693594
3. ATLAS Collaboration (2022) Direct constraint on the Higgs-charm coupling from a search for Higgs boson decays into charm quarks with the ATLAS detector. arXiv:2201.11428. Submitted to EPJC
4. ATLAS Collaboration (2021) Measurements of WH and ZH production in the $H \rightarrow b\bar{b}$ decay channel in pp collisions at 13 TeV with the ATLAS detector. Eur Phys J C 81:17.https://doi.org/10.1140/epjc/s10052-020-08677-2. arXiv:2007.02873
5. Zurbano Fernandez I et al High-luminosity large hadron collider (HL-LHC): technical design report. https://doi.org/10.23731/CYRM-2020-0010
6. ATLAS Collaboration (2018) Search for the decay of the Higgs boson to charm quarks with the ATLAS experiment. Phys Rev Lett 120:211802. https://doi.org/10.1103/PhysRevLett.120.211802. arXiv:1802.04329
7. ATLAS Collaboration (2018) Prospects for $H \rightarrow c\bar{c}$ using charm tagging with the ATLAS experiment at the HL-LHC. ATLAS PUB Note ATL-PHYS-PUB-2018-016
8. ATLAS Collaboration (2021) Extrapolation of ATLAS sensitivity to $H \rightarrow b\bar{b}$ and $H \rightarrow c\bar{c}$ decays in VH production at the HL-LHC. ATLAS PUB Note ATL-PHYS-PUB-2021-039
9. OPAL Collaboration (2000) A Measurement of the rate of charm production in W decays. Phys Lett B 490:71–86. https://doi.org/10.1016/S0370-2693(00)00971-0. arXiv:hep-ex/0009020
10. ALEPH Collaboration (1999) A direct measurement of |V(cs)| in hadronic W decays using a charm tag. Phys Lett B 465:349–362. https://doi.org/10.1016/S0370-2693(99)01088-6

Chapter 11
Conclusion

In this thesis, direct searches for the Standard Model (SM) Higgs boson, Z boson, and W boson decaying to charm (c) quark(s) have been presented. The searches use proton-proton (pp) collision data at a centre-of-mass energy of 13 TeV collected with the CMS detector in 2016–2018 and corresponding to an integrated luminosity of 138 fb^{-1}. The search is carried out in two topologies: the resolved-jet topology and the merged-jet topology. The former is the focus of the thesis and employs a new calibration method for c quark identification, which has also been discussed here in detail. The Higgs boson search (VHcc) is carried out in the vector-boson (V) associated production of the Higgs boson, where the V decays leptonically. The Z boson (VZcc) and W boson (VWcq) searches are carried out in diboson events with a final state similar to that of the VHcc search.

A statistical combination of the results from the resolved-jet approach, that has been discussed in detail, with those from the complementary merged-jet approach, allows us to set the strongest observed (expected) direct upper limits on the VHcc process: 14 (7.6) times the SM expectation at the 95% Confidence Level (CL). The corresponding observed (expected) best-fit signal strength is $7.7^{+3.8}_{-3.5}$ ($1.0^{+3.7}_{-3.4}$), compatible with the SM expectation within 2σ. The sensitivity achieved in the presented search with 138 fb^{-1} pp collision data is comparable with the past expectations of sensitivities achievable with 3000 fb^{-1} of collision data to be available in the future with the HL-LHC upgrade.

In a similar fashion, the combination of the two topologies yields an observed (expected) VZcc significance of 5.7σ (5.9σ) over the background-only prediction, with a corresponding best-fit signal strength of $1.01^{+0.23}_{-0.21}$ ($1.00^{+0.22}_{-0.20}$) at the 95% CL. This corresponds to an observation of the VZcc process with 5.7σ significance. The resolved-jet topology alone is suitable for the VWcq search and results in an observed (expected) VWcq significance of 5.6σ (5.7σ) over the background-only prediction, with a corresponding best-fit signal strength of $1.02^{+0.20}_{-0.19}$ ($1.00^{+0.20}_{-0.18}$). This corresponds to an observation of the VWcq process with 5.6σ significance.

S. Mondal, *Charming Decays of the Higgs, Z, and W Bosons*, Springer Theses,
https://doi.org/10.1007/978-3-031-84677-9_11

The VHcc search is motivated by the fact that it allows for a direct measurement of the Higgs-charm coupling, which in turn helps verify or exclude Beyond the SM (BSM) couplings of the Higgs boson to second-generation quarks. The current search excludes BSM scenarios that predict more than 14 times the SM production rate of the VHcc process, at the 95% CL. The analogous VZcc search serves as a validation of the novel tools and analysis techniques used in the VHcc search. The VZcc and VWcq searches also provide first observations of the respective processes at hadron collider experiments, attesting to the unprecedented sensitivities currently achievable at the CMS experiment, in general, and advancements in charm tagging capabilities, in particular.

The sensitivity of the VHcc search is limited by the statistical uncertainties in the selected events. Stronger limits can be obtained in the future by combining the results of the present search with results from the data collected in the Run-3 of the LHC. Complementary constraints on the Higgs-charm coupling can also be obtained through searches of the Higgs boson produced in association with a charm quark. Such future searches will also benefit from more advanced flavour tagging algorithms that are currently under development. The measurement of the Higgs-charm coupling will remain an important goal of the LHC physics programme in the coming decade.

This thesis also presents a novel method to calibrate the full differential shape of the discriminator distributions used for c jet identification at CMS. The method uses three different sets of event selection criteria, targeting topologies enriched in W+c, top quark pairs, and Drell–Yan+jet events. By employing an iterative fitting approach in each of these three regions, scale factors (SFs) are derived to match the simulated discriminator distributions to those observed in data. The calibration of the full differential discriminator shape allows the use of the c tagging discriminator scores as inputs to multivariate techniques (based on machine learning). The shape calibration extends the use of c tagging algorithms beyond the application of discrete working points, and facilitates more advanced uses for c jet identification in physics analyses. Furthermore, this approach is extended to utilise calibrated c tagger scores of two jets to perform a correlated charm-strange (cs) tagging. This approach is successfully implemented in the context of the VWcq search in this thesis, and can be extended to hypothetical charged Higgs boson ($H^{\pm} \rightarrow$ cs) searches and measurement of the charmed W decay fraction in the future.

Appendix A
Charm Tagging

Normalised 2D distributions of jets with their CvsL and CvsB discriminators on the x and y axes, respectively, are shown in Fig. 6.6 in Sect. 6 of the thesis. These plots are remade by plotting all three flavours in one plot and are shown in Fig. A.1.

S. Mondal, *Charming Decays of the Higgs, Z, and W Bosons*, Springer Theses,
https://doi.org/10.1007/978-3-031-84677-9

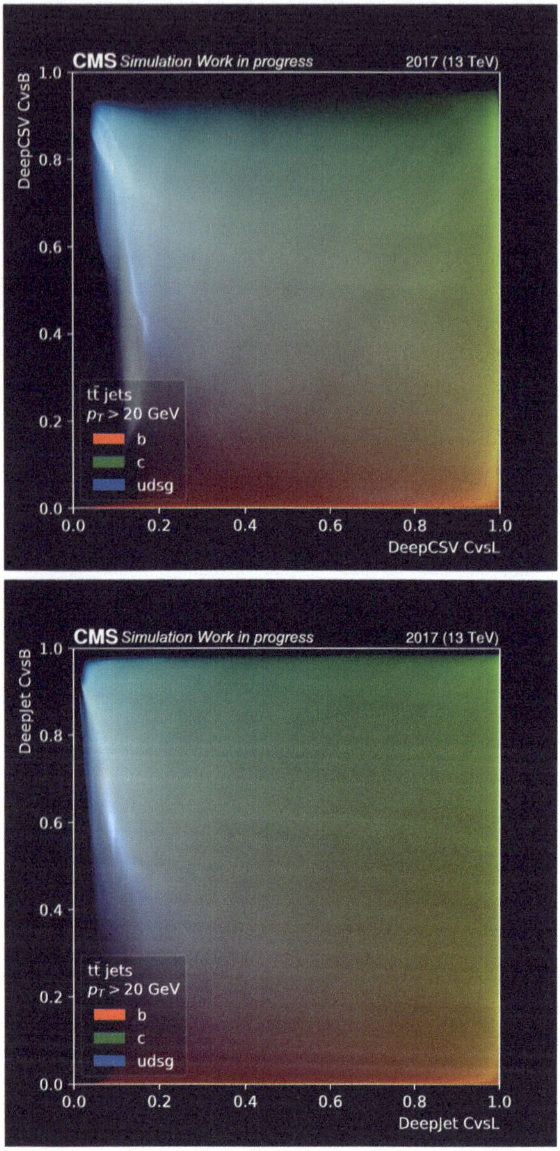

Fig. A.1 Jets of various flavours in the CvsL–CvsB plane for the DeepCSV- (upper) and DeepJet- (lower) based charm taggers taken from a hadronic $t\bar{t}$ sample simulated with 2017 data-taking conditions. Each pixel in the plane is assigned a colour using the RGB scheme, such that the values of red, green, and blue components each depends on the relative proportions of b, c, and udsg jets, respectively, in that point on the plane

Appendix B
Simulated Samples

This appendix summarises the simulated samples used in the VHcc, VZcc, and VWcq analyses, in Tables B.1, B.2, B.3, B.4, B.5, B.6, and B.7. The description of the samples can be found in Sect. 9.2.

Table B.1 The simulated VHcc signal samples with their cross sections (XSec) used in the analyses

Process	Samples	Generator	Used in analysis?			XSec (pb)
			VHcc	VZcc	VWcq	
WH(cc)	$W^+(l^+\nu)$+H($c\bar{c}$)	POWHEG	✓	✓		0.008540
	$W^-(l^-\nu)$+H($c\bar{c}$)		✓	✓		0.005411
ZH(cc)	$q\bar{q}\rightarrow$Z(l^+l^-)+H($c\bar{c}$)	POWHEG	✓	✓		0.002342
	gg $\rightarrow$Z(l^+l^-)+H($c\bar{c}$)		✓	✓		0.000357
	$q\bar{q}\rightarrow$Z($\nu\nu$)+H($c\bar{c}$)		✓	✓		0.004628
	gg $\rightarrow$Z($\nu\nu$)+H($c\bar{c}$)		✓	✓		0.000713

Table B.2 The simulated VHbb background samples with their cross sections (XSec) used in the analyses

Process	Samples	Generator	Used in analysis?			XSec (pb)
			VHcc	VZcc	VWcq	
WH(bb)	$W^+(l^+\nu)$+H($b\bar{b}$)	POWHEG	✓	✓		0.17202
	$W^-(l^-\nu)$+H($b\bar{b}$)		✓	✓		0.10899
ZH(bb)	$q\bar{q}\rightarrow$Z(l^+l^-)+H($b\bar{b}$)	POWHEG	✓	✓		0.04718
	gg $\rightarrow$Z(l^+l^-)+H($b\bar{b}$)		✓	✓		0.00720
	$q\bar{q}\rightarrow$Z($\nu\nu$)+H($b\bar{b}$)		✓	✓		0.09322
	gg $\rightarrow$Z($\nu\nu$)+H($b\bar{b}$)		✓	✓		0.01437

S. Mondal, *Charming Decays of the Higgs, Z, and W Bosons*, Springer Theses,
https://doi.org/10.1007/978-3-031-84677-9

Table B.3 The simulated diboson samples with their cross sections (XSec) used with datasets corresponding to the 2016 data-taking conditions. The VWcq signal cross sections have been reported at NLO accuracy, while others have been evaluated at LO accuracy and rescaled to NLO by multiplying a so-called k-factor. WW cross sections have been rescaled to NNLO. aMC@NLO refers to MadGraph5_aMC@NLOrun in NLO mode

Process	Samples	Generator	Signal of...	XSec (pb)	k-factor
WW	W($l\nu$)+W($l\nu$)	Powheg	–	12.178	1.227
	W($l\nu$)+W(qq)		VWcq	43.547	–
WZ	W($l\nu$)+Z($q\bar{q}$)	aMC@NLO	VZcc	10.71	1.056
	W($l\nu$)+Z($\nu\nu$)		–	3.06	1.056
	W(qq)+Z(ll)		VWcq	5.595	–
	W(qq)+Z($\nu\nu$)		VWcq	6.317	–
	W($l\nu$)+Z(ll)		–	4.692	1.056
ZZ	Z(ll)+Z($\nu\nu$)	Powheg	–	0.564	1.208
	Z(ll)+Z($q\bar{q}$)	aMC@NLO	VZcc	3.22	1.208
	Z($q\bar{q}$)+Z($\nu\nu$)	aMC@NLO	VZcc	4.04	1.208
	Z(ll)+Z(ll)	Powheg	–	1.256	1.208

Table B.4 The simulated diboson samples with their cross sections (XSec) used with datasets corresponding to the 2017 and 2018 data-taking conditions. Cross sections have been evaluated at LO accuracy and rescaled to NLO by multiplying a so-called k-factor. aMC@NLO refers to MadGraph5_aMC@NLO run in NLO mode

Process	Samples	Generator	Signal of...	XSec (pb)	k-factor
WW	W($l\nu$)+W($l\nu$)	Powheg	–	11.08	1.162
	W($l\nu$)+W(qq)		VWcq	45.99	1.162
WZ	W($l\nu$)+Z($q\bar{q}$)	aMC@NLO	VZcc	11.66	0.978
	W($l\nu$)+Z($\nu\nu$)		–	3.294	0.978
	W(qq)+Z(ll)		VWcq	6.351	0.978
	W(qq)+Z($\nu\nu$)		VWcq	6.317	0.978
	W($l\nu$)+Z(ll)		–	5.062	0.978
ZZ	Z(ll)+Z($\nu\nu$)	Powheg	–	0.6008	1.134
	Z(ll)+Z($q\bar{q}$)	aMC@NLO	VZcc	3.693	1.134
	Z($q\bar{q}$)+Z($\nu\nu$)	aMC@NLO	VZcc	4.554	1.134
	Z(ll)+Z(ll)	Powheg	–	1.325	1.134

Table B.5 The simulated $t\bar{t}$ and ST samples with their cross sections (XSec)

Process	Samples	Generator	Year(s)	XSec (pb)
$t\bar{t}$	Dilep. $t\bar{t}$	POWHEG	All	88.29
	Semilep. $t\bar{t}$			365.34
	Hadronic $t\bar{t}$			377.96
SingleTop	s-channel	AMC@NLO	2016	3.36
	s-channel t	POWHEG	2017–18	2.0688
	s-channel $\bar{t}$		2017–18	1.2934
	t-channel t		All	136.02
	t-channel $\bar{t}$		All	80.95
	tW		All	35.85
	$\bar{t}$W		All	35.85

Table B.6 The simulated V+jets samples with their cross sections (XSec) used with datasets corresponding to the 2016 data-taking conditions. Cross sections have been evaluated at NNLO accuracy

Process	V p_T (GeV)	XSec (pb)
DY(ll)+jets	Inclusive	6508.0
	50–100	363.81
	100–250	84.015
	250–400	3.2283
	400–650	0.43604
	650–∞	0.04098
W($l\nu$)+jets	50–100	3298.37
	100–250	689.75
	250–400	24.507
	400–600	3.1101
	600–∞	0.46832
Z($\nu\nu$)+jets	50–100	688.92
	100–250	159.95
	250–400	6.0386
	400–650	0.81127
	650–∞	0.075855

Table B.7 The simulated V+jets samples with their cross sections (XSec) used with datasets corresponding to the 2017 and 2018 data-taking conditions. Cross sections have been evaluated at NNLO accuracy. Number of additional partons at the generator level is indicated in the "# add. partons" column. The W($l\nu$)+jets with V $p_T \in [50, 150]$ samples have been used with an addition criterion of V $p_T < 100$ GeV at the generator level

Process	# add. partons	V p_T (GeV)	XSec (pb)
DY(ll)+jets	0–4	Inclusive	6508.0
	1	50–150	316.6
		150–250	9.543
		250–400	1.098
		400–∞	0.1193
	2	50–150	169.6
		150–250	15.65
		250–400	2.737
		400–∞	0.447
W($l\nu$)+jets	1	50–150	2661.0
		100–150	286.1
		150–250	71.9
		250–400	8.05
		400–∞	0.885
	2	50–150	1331.0
		100–150	277.7
		150–250	105.9
		250–400	18.67
		400–∞	3.037
Z($\nu\nu$)+jets	1	50–150	596.3
		150–250	17.98
		250–400	2.057
		400–∞	0.224
	2	50–150	325.7
		150–250	29.76
		250–400	5.166
		400–∞	0.8457

The manufacturer's authorised representative in the EU is Springer
Nature Customer Service Centre GmbH, Europaplatz 3, 69115 Heidelberg,
Germany. If you have any concerns regarding our products, please
contact ProductSafety@springernature.com

Printed and bound by CPI Group (UK) Ltd, Croydon, CR0 4YY
30/09/2026
02225577-0013